Using *MultiSIM*®:
Troubleshooting DC/AC Circuits

Using MultiSIM®:
Troubleshooting DC/AC Circuits

Second Edition

John Reeder

Merced College
Merced, CA

THOMSON

DELMAR LEARNING

Australia   Canada   Mexico   Singapore   Spain   United Kingdom   United States

THOMSON

DELMAR LEARNING

Using *MultiSIM*®: Troubleshooting DC/AC Circuits, Secnd Edition
John Reeder

Vice President, Technology and Trades SBU:
Alar Elken

Editorial Director:
Sandy Clark

Acquisitions Editor:
Steve Helba

Developmental Editor:
Michelle Ruelos Cannistraci

Marketing Director:
Dave Garza

Marketing Coordinator:
Casey Bruno

Production Director:
Mary Ellen Black

Production Manager:
Larry Main

Senior Project Editor:
Christopher Chien

Art/Design Coordinator:
Francis Hogan

Editorial Assistant:
Dawn Daugherty

ISBN: 1-4018-5264-8

NOTICE TO THE READER

Publisher does not warrant or guarantee any of the products described herein or perform any independent analysis in connection with any of the product information contained herein. Publisher does not assume, and expressly disclaims, any obligation to obtain and include information other than that provided to it by the manufacturer.

The reader is expressly warned to consider and adopt all safety precautions that might be indicated by the activities herein and to avoid all potential hazards. By following the instructions contained herein, the reader willingly assumes all risks in connection with such instructions.

The publisher makes no representation or warranties of any kind, including but not limited to, the warranties of fitness for particular purpose or merchantability, nor are any such representations implied with respect to the material set forth herein, and the publisher takes no responsibility with respect to such material. The publisher shall not be liable for any special, consequential, or exemplary damages resulting, in whole or part, from the readers' use of, or reliance upon, this material.

Contents

Using MultiSIM®:
Troubleshooting DC/AC Circuits

Second Edition

John Reeder

Merced College
Merced, CA

THOMSON

DELMAR LEARNING ™

Australia   Canada   Mexico   Singapore   Spain   United Kingdom   United States

THOMSON

DELMAR LEARNING

Using *MultiSIM*®: Troubleshooting DC/AC Circuits, Secnd Edition
John Reeder

Vice President, Technology and Trades SBU:
Alar Elkèn

Editorial Director:
Sandy Clark

Acquisitions Editor:
Steve Helba

Developmental Editor:
Michelle Ruelos Cannistraci

Marketing Director:
Dave Garza

Marketing Coordinator:
Casey Bruno

Production Director:
Mary Ellen Black

Production Manager:
Larry Main

Senior Project Editor:
Christopher Chien

Art/Design Coordinator:
Francis Hogan

Editorial Assistant:
Dawn Daugherty

ISBN: 1-4018-5264-8

NOTICE TO THE READER

Preface

Approach

This workbook is designed to teach the student how to virtually measure and troubleshoot electronic circuits created within the MultiSIM (EWB) software environment and to reinforce DC/AC theory learned in the classroom. The computer and the computer monitor become the electronics workbench. Students using this manual must have the latest version of MultiSIM (currently version 7) installed on their computer to be able to operate the provided software projects. Earlier versions of MultiSIM or Electronics Workbench will not operate the Version 7 files on the accompanying CD. These software projects are stored on the CD that accompanies this text.

An advantage to the virtual laboratory approach to electronics is the low cost of the software package in comparison to the expenses required to establish an electronics laboratory with all of the necessary test equipment and the related facility costs. Virtual electronic circuits can be modified easily on the monitor screen, and circuit analysis is also easily obtained as circuits are modified.

Circuit troubleshooting is an integral component of the software package. It is relatively easy for the instructor or textbook author to install faults such as shorts, leakage, and opens into the circuit for the electronics student to locate. The troubleshooting exercises will provide the student with the confidence and skills necessary to troubleshoot circuits constructed on the electronics laboratory workbench.

System requirements

Pentium II+
Windows® 98/NT 4/XP
64MB RAM (128MB RAM recommended)
100–250 MB hard disk space (min.)
CD-ROM drive
800×600 minimum screen resolution

Organization

Each chapter in this workbook attempts to sequentially follow the material found in most textbooks teaching DC/AC theory. **Activity** sections that break the overall subject of the chapter down into smaller blocks are located in each chapter. The individual software projects are related to subtopics within the larger topic.

Within each activity, **circuit files** progressively provide individual projects related to the subject material of the chapter. Most subjects are touched upon as progress is made through the projects. A final circuit file for many of the activity sections is a **troubleshooting** problem.

Circuit Files

The CD that accompanies the text contains all of the circuit files in this book. They are pre-built and ready to use with MultiSIM 7. There are over 550 circuits available. Twenty-five percent of the circuits are enabled with the accompanying textbook edition of MultiSIM. All circuits can be opened using MultiSIM 7.

Circuit files follow the DOS system of nomenclature. There are a maximum of eight digits in each circuit name. The first two numbers of the circuit file name represents the chapter in the book followed by a dash. After the dash, the following numbers and letters represent the activity within the chapter. The letter, and occasionally a letter followed by a number, represents the sequence of events within an activity.

Instrumentation

Test instruments are accessed through the **Instruments** button on the Design Bar above the circuit workspace. A left-click on this button will cause the instruments toolbar to appear at the bottom of the screen. You will use the digital multimeters (DMMs), the oscilloscopes, and the Bode plotter with this workbook. The DMMs and the oscilloscopes are common instruments found on the test bench. The Bode plotter is a virtual instrument found only in a software program; it is not a real-world instrument. The Bode graphic plot, a diagram of voltage amplitude of a circuit in reference to the circuit frequency, is commonly used in electronics to better understand circuit operation as frequency changes. The Bode plotter used in MultiSIM displays the Bode plot of a circuit as if it were a piece of a real test instrument; it is similar to a spectrum analyzer, an advanced piece of test equipment found in some electronics test labs. This latest version of MultiSIM has added Agilent (virtual) instruments to the Instruments toolbar.

Additional Resources

Thomson Delmar Learning provides support for electronics instructors on their Web site at www.electronictech.com. Answers to questions and problems in the text will be provided on an instructor's CD (ISBN: 1-4018-5255-6).

Interactive Image Technologies, the producers of MultiSIM, can be contacted at (416) 977-5550 for sales or technical support. Their Web page can be located at www.electronicsworkbench.com. They can also be reached through their sales phone number, 1-800-263-5552.

About the Author

John Reeder, A.A., B.A., M.S., is a retired electronics instructor from Merced College, after teaching electronics at the high school, community college, and university level for the past 20 years. At present, he lectures in electronics and advanced automation at California State University, Fresno. Prior to becoming a teacher, he worked for 28 years in the electronics and electrical industries as technician, electrician, and engineer. He has been using Electronics Workbench® and MultiSIM for many years as a supplement to help students gain a better understanding of their electronics material. At Merced College, he developed course curriculum using EWB/MultiSIM as the software core of the overall electronics program.

The author, as an instructor for this material, welcomes input regarding the content of this book. Please address comments, questions, and suggestions to:

ms7@jreeder.info or jreeder95348@yahoo.com

Information about the textbook and MultiSIM will be posted on the author's web site at:

http://www.jreeder.info.

Acknowledgments

I would like to express my grateful appreciation to Michelle Ruelos Cannistraci at Thomson Delmar Learning for patiently working with me and helping me over the hurdles of this first revision of my book.

I would like to express my grateful appreciation to the software team and the technical staff at Interactive Image Texhnologies who helped me through the difficult moments when the software wasn't responding to my efforts. All worked out well and the result is this finished product.

I also want to thank my partners in electronics instruction at Merced College, Bill Walls and Eugen Constantinescu, who worked with me through the years as I wrote and revised this text. I also want to thank my department head at CSU Fresno, Dr. Tony Au, who gave me the green light to continue using my two MultiSIM texts in my classes at the university.

In addition, I want to thank my students at Merced College and California State University, Fresno, who provided the continuing impetus to keep the book current and useful in their studies. Their enthusiasm over the book and the MultiSIM projects has been wonderful.

Last, but not least, I have to acknowledge the most important member of my team, my wife Barbara, who put up with the long hours spent on the computer this past year. She always provided the right touch, and a cup of coffee at the right time, to keep me going in the critical moments.

1. Introduction to MultiSIM®: The Electronics Lab in the Computer

References:

Electronics Workbench®, *MultiSIM* Version 7

Electronics Workbench®, *MultiSIM* Version 7 User's Guide

Objectives   After completing the chapter the student should be able to:

- Open and use the MultiSIM program
- Open MultiSIM component and instrument libraries
- Construct basic electronics circuits in MultiSIM

Working within the Microsoft Windows® Environment

The MultiSIM "laboratory simulation" software uses the Microsoft Windows® 98/NT4/2000XP operating systems that are installed on many desktop computers (Pentium II or higher) found in homes, schools, and industry. On the workspace of the computer screen that you are using there should be an icon that will activate the installed MultiSIM software program. When this icon is activated, the MultiSIM program will load and await your command: a virtual electronics laboratory at your disposal.

Opening and Using MultiSIM

Introducing *MultiSIM*

When the program is first activated, the display presents a blank workspace entitled "Circuit 1" (Figure 1-1). This window can be used to construct a circuit of choice, to display a "pre-built" circuit found within the program file system, or to display a circuit file found in another location such as a CD. When looking for circuits within the file system or another location, start by left-clicking on **File** and **Open** with the mouse as with any Windows® program. Then open the file folder where the circuit file is located, such as the hard drive file folder or the CD that accompanies this text. The file system used within MultiSIM is the same as

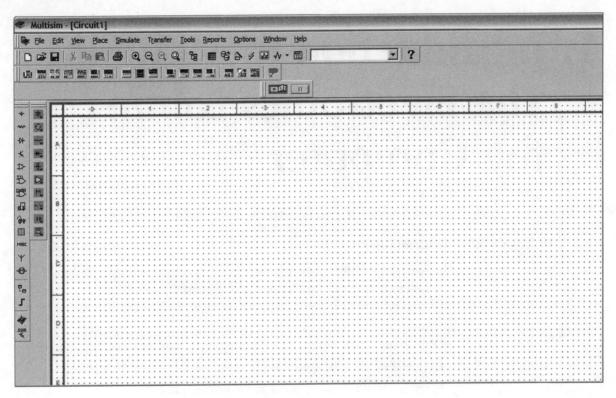

Figure 1-1 MultiSIM Opening Screen

the Microsoft Windows® file system and can be manipulated using the same methods and techniques as File Manager or Explorer. If the project files are installed on the computer hard drive, they should be located in the MultiSIM folder DC_AC Files as displayed in Figure 1-2.

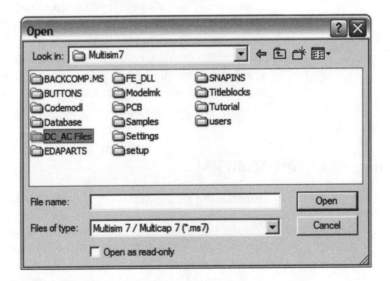

Figure 1-2 File Menu for Circuit Files on the Hard Drive

MultiSIM is best defined as an electronics workbench in a virtual laboratory. Practically every circuit studied in lower level electronics courses can be constructed and tested using this powerful software application on a computer. In a real laboratory environment, MultiSIM can be used to verify and compare concepts concerning electronic circuits constructed on the workbench. This workbook is primarily designed to reinforce subject material covered in any electronics textbook used for DC/AC coursework in secondary and post-secondary education.

Component and Instrument Libraries

Launch MultiSIM and observe the blank "Circuit 1" window. The **Component** toolbar is located to the far left of the circuit window. This toolbar contains thirteen component parts bins. To the right of this toolbar there is a shorter toolbar entitled **Virtual Toolbar**. Both toolbars are displayed in Figure 1-3 in the horizontal position rather than their normal vertical position on the left of the screen.

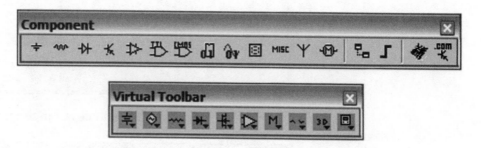

Figure 1-3 The MultiSIM Component and Virtual Toolbar

The MultiSIM user can access the individual libraries with the mouse using "left-click" techniques; left-click on the proper library, left-click on the desired component, left-click on the **OK** button, and then, as the component follows the mouse onto the workspace, you left-click it into place. Each of these libraries can be identified, regarding their contents, by resting the mouse pointer on a specific library and waiting for the drop-down **Tool Tip** to appear. While proceeding through this text, the primary component libraries that will be used are **Sources** (Figure 1-4), **Basic** (Figure 1-5), **Indicators** (Figure 1-6), and **Miscellaneous** (Figure 1-7). The **Instruments** library (Figure 1-8) is located above the workspace.

Using the mouse, access each of the component libraries and observe their contents—contents that will become very familiar while continuing on in the study of electronics technology with the help of MultiSIM. Locate the position of **Sources**, **Basic**, **Indicators**, **Miscellaneous**, and finally the **Instruments** library. Use the drop-down **Tool Tip** to verify the library locations.

MultiSIM contains typical test equipment that is found in an electronics laboratory (except that it is virtual test equipment). This inventory of test equipment includes a multimeter, a function generator, an oscilloscope, and a Bode plotter that will be used in this text. There is also other test equipment that will not be used at this point in your study of electronics. In addition, within the **Indicators** library, there are digital voltmeters, digital ammeters, voltage probes, and other indicating devices. These devices from the **Indicators** library may be used as many times as necessary in a circuit, as may the instruments within the **Instruments** library. New

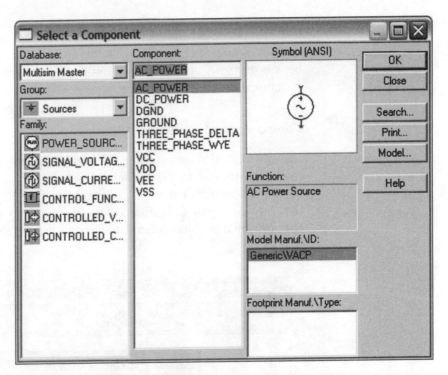

Figure 1-4   The Sources Component Library

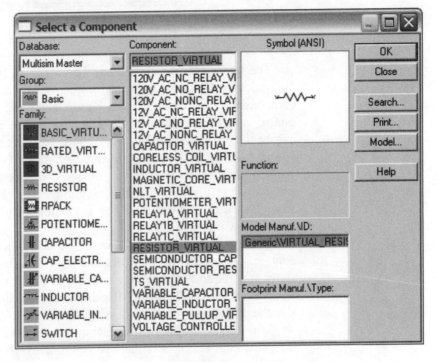

Figure 1-5   The Basic Component Library

to the latest version of MultiSIM is virtual Agilent® test equipment consisting of a multimeter, a function generator, and an oscilloscope.

Agilent® is a standard brand of test equipment that is frequently found in the laboratories of the electronics industry. There are other brands that are as good

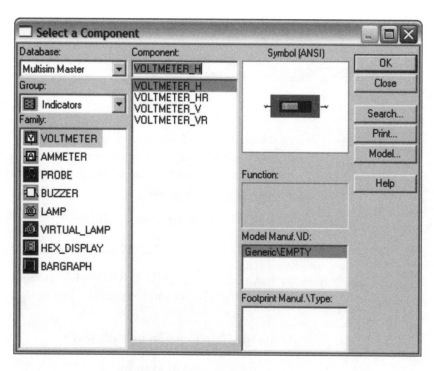

Figure 1-6 The Indicators Library

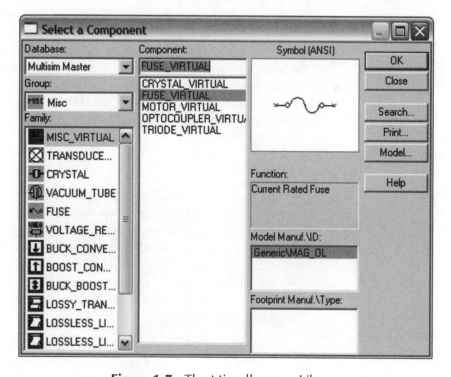

Figure 1-7 The Miscellaneous Library

and found as frequently, but Agilent test equipment is a good choice for you to learn to use. If you wish to learn more about the Agilent test equipment, you can download the User Guides for each of the Agilent test instruments (Adobe

Figure 1-8   The Instruments Toolbar

Acrobat Reader® is required to read these files) from the MultiSIM software site at *http://www.electronicsworkbench.com* or the Agilent library Web site at *http://we.home.agilent.com/cgi-bin/bvpub/agilent/library/cp_library.jsp*. These test instruments are:

1. 34401-90004: HP34001A Multimeter

2. 54622-97036: 54622D Mixed Signal Oscilloscope

3. 6C0633120A_USERSGUIDE_ENGLISH: 33120A 15MHZ Function Arbitrary Waveform Generator.

Using the Help function in *MultiSIM*

Any time you need to know more about a component or an instrument, the first step is to drag the component or instrument out onto the workspace. Then use one of several possible methods to obtain the desired knowledge. With the first method, point the mouse on the component or instrument, right-click to bring up a menu, and then left-click on **Help** at the bottom of the menu (or use the F1 key). A second method is to left click on the large **?**, located on the top toolbar and type in a term related to your component or test instrument to obtain information. Either method will bring up help (information) about the component or instrument. You can also use the large **?** or **Help** menu at the top of the screen to access additional information about the MultiSIM software package.

Saving your work

When you have completed your project or the work period is nearing an end, you can save your work to the computer hard drive or a floppy disk. To save your file, open the drop-down **File** menu at the upper right of the taskbar. Left click on *Save As...* and the menu shown in Figure 1-9 will appear. Double left-click on the **Users** folder and the menu of Figure 1-10 will open. Save your circuit with a name that you can remember (others may be using the same computer at another time). You can also save the file on the A drive (floppy disk drive) and in this location under "whatever_name.ms7."

Opening a Circuit File and Building Your First Circuit

1. Now is the time to build your first circuit. The first step is to activate the MultiSIM software. You are going to duplicate an already built circuit from the Chapter 1 circuit file folder (Figure 1-11).

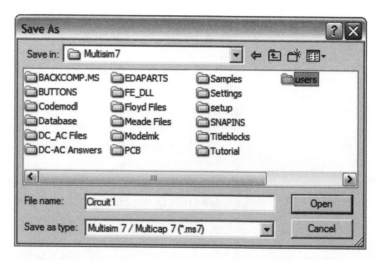

Figure 1-9   Saving a Circuit File to the Hard Drive

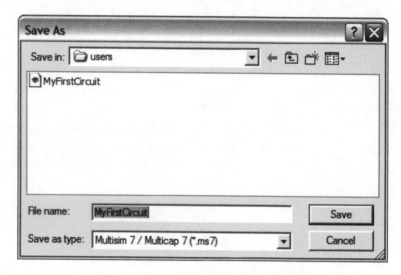

Figure 1-10   Saving to the User's Folder

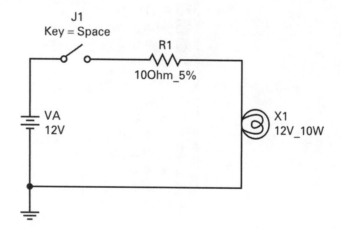

Figure 1-11   A Circuit to Duplicate

2. Go to the file drop-down menu at the top left of the screen and open circuit file **01-01**. Notice that there is a pre-built demonstration circuit for you to copy. You are to duplicate this pre-built circuit according to the following steps:

    a. Open the Sources library and left-click on DC_POWER. Left-click on the OK button. The menu should disappear and the battery symbol (V2) should appear on the workspace; move it with the mouse to a good working location.

    b. Open the **Sources** library and, using the same procedure, obtain the **GROUND** symbol and install it below the battery symbol in a manner similar to the demonstration circuit.

    c. Move your mouse to the bottom of the battery symbol and, when the crosshair **(+)** appears, left-click the mouse. The crosshair replaces the mouse arrow. Now move the crosshair to the top of the ground symbol and, when the two touch, left-click again. You should now have a wire (line) from the bottom of the battery symbol to the top of the ground symbol. You will follow this procedure throughout the text when working with circuits.

    d. Open the **Basic** library, select **Resistor** (fourth from top), and obtain the default 1.0 ohm resistor **(R2)**. Place it on the workspace, leaving room for a switch.

    e. Return to the **Basic** library and select a SPST switch **(J2)**. Install it to the left of the resistor. Now install circuit wires (lines) from the battery to the switch and from the switch to the resistor. See Figure 1-12 for switch selection.

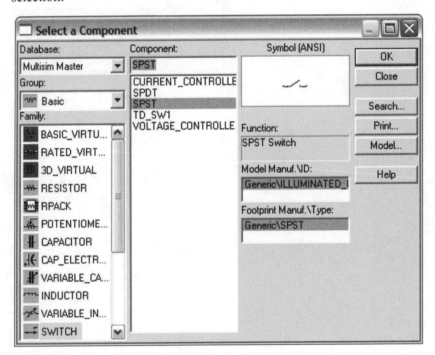

Figure 1-12   Selecting a Switch from the Basic Library

f. Open the **Indicator** library, select **Lamp** and then choose the 12V_10W lamp. After placing the lamp **(X2)** to the right of the battery symbol in a manner similar to the demonstration circuit, right-click on this lamp and select **90 Clockwise** from the menu. Connect (wire) the lamp to the resistor and the wire between the bottom of the power source (the battery) and the ground symbol.

g. Activate the circuit by clicking on the **Run/stop simulation** switch at the top of the screen (or toggle **F5** on the keyboard). Turn the circuit on (including the demo circuit) with the **Spacebar** on the keyboard. The lamp in both circuits should indicate the **ON** condition.

h. Congratulations, you have completed your first circuit. Now the real work (or fun, depending on how you look at it) starts.

Regarding Answers to Questions

Measured indicates that a MultiSIM test instrument(s) was used to determine the value of an electrical quantity.

Displayed indicates that the value was determined from the workspace (i.e., a component value or a meter reading).

Calculated means that the answer was determined by means of one of the laws or rules of electronics such as Ohm's Law, using known values from the schematic or test instruments. For example, in the circuit file for Chapter 3, Activity 3.1, Step 2 (03-01a), the value for V_A is displayed near the power source, whereas the ammeter measures the value of I_T.

2. Introduction to Electricity and Electronics: Electrical Quantities and Components

References

Electronics Workbench®, MultiSIM Version 7

Electronics Workbench®, MultiSIM Version 7 User's Guide

Objectives   After completing this chapter the student should be able to:

- Recognize basic MultiSIM component symbols used for resistors, DC power sources (batteries), inductors, capacitors, diodes, transistors, transformers, switches, fuses, and connection points
- Activate pre-built circuits
- Recognize and use MultiSIM components
- Construct a basic circuit and observe the effects of DC current flow
- Use virtual digital multimeters (DMM) to measure electrical quantities
- Determine resistor values using resistor color codes
- Determine and vary the resistance of potentiometers in MultiSIM
- Validate and use Ohm's law
- Calculate power consumption
- Develop and use troubleshooting skills to locate faulty resistors

Introduction

The knowledge of electronics begins with a thorough understanding of the foundational phenomena of charge, voltage, current, resistance, capacitance, inductance, and power. All of electronics hangs on these root concepts. Electronics circuits exhibit all of these basic phenomena in varying degrees and are dependent upon the value of the circuit components and the amount of voltage applied to the circuit by the voltage source.

Electronic components are the basic building blocks of a circuit. If you look at the resistor, as an example, you discover that it is a component that exhibits certain electrical characteristics with the amount of resistance being identified through a color code system (in the case of most resistors). This you will study. Other components will be studied and explained as you progress through the text.

The circuit diagram, or schematic diagram, can be defined as a symbolic representation of an electrical/electronic circuit. Through the schematic diagram, the physical configuration of a circuit can be determined and then constructed according to the diagram, subject to physical constraints.

Activity 2.1: Electronic Components and Power Sources

Electronic components are the manufactured parts used to construct electronics circuits. These components range from small parts, such as transistors, to very complex electronic assemblies. The electronic assemblies may, within themselves, contain many individual components and integrated circuits that became a part of these more complex assemblies in the manufacturing process. Various small components and even very complex manufactured circuits and assemblies become the basic building blocks of ever more complex electronic equipment and systems. MultiSIM has the feature of virtual components as well as fixed (menu selected) components. Virtual components give the user the opportunity to easily change the value of a component rather than using a fixed menu. Resistors have the additional virtual capability of setting the wattage rating of a resistor. This wattage rating causes the resistor to open (burn up) if the wattage rating is exceeded.

1. **Resistors** are components that are used to oppose or limit current flow in electronic circuits. On the MultiSIM workspace, in close proximity to the resistor symbol, a component designator, a resistance value, and special labels may be displayed. In Figure 2-1, for example, the resistor component symbols have the component designators (R1, R2, and R3) and the resistance value (1K) near the components. R1 and R2 are basic resistors, and R3 is a variable resistor, also known as a potentiometer. The potentiometer has many applications and is widely used as a volume control in audio equipment.

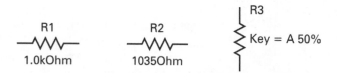

Figure 2-1   MultiSIM Resistive Symbols

2. Open circuit file **02-01a**. It contains resistive components from the MultiSIM **Basic** menu (R1 through R4) and a pictorial of R1. A double left-click on any of the components brings up the **Component Properties** menu. R1 has a fixed value, and if you wish to change its value, you must click on **Replace** and select a new value from the menu. To change the value of R2 or R3, click on the **Value** tab and change the value. R3 offers the additional option of selecting a wattage value that simulates real circuit action by causing the resistor to "burn open" if the resistor wattage rating is exceeded in a circuit. Circuit 1 demonstrates this action. Activate the circuit using the **F5** key, close the switch with the **Spacebar** and wait a few seconds for R5 to open. A single right-click on a component brings up another menu with additional

options such as rotate (try rotating one of the components). These methods of changing component settings and values are used throughout the MultiSIM program.

3. Now, open the **Basic** menu on the toolbar (the icon with the resistor on it in the first column), left-click on **Basic_Virtual**, next, left-click on **Resistor_Virtual**, and finally, left-click on **OK**. A virtual resistor will appear on your workspace. Double left-click on the resistor and the component menu will appear. Change the value of the resistor to 10k Ohm. Select the **Display** tab, check off the check mark on **Use Schematic Global Setting** (notice that **labels**, **values**, **reference ID**, and **Attributes** are checked; these can be turned off). Select the **Label** tab and change the **Reference ID** to R7. Add a **Label** of your choice (for example, My_Resistor), and finally, left-click on **OK**. The designations around the resistor should have changed to 10 k Ohm, R7, and the label of your choice.

4. **DC Power Sources (batteries)** provide DC power for the circuits and are represented in MultiSIM by the battery symbol. The DC power sources (power supplies) may also have the component designator, a DC voltage value, and special labels displayed near the component. In Figure 2-2, the component designator **(V1)** and the DC voltage value **(12V)** are shown near the component symbol. There are several other types of power source symbols in MultiSIM that will be introduced later.

Figure 2-2   The MultiSIM Battery Symbol

5. Open circuit file **02-01b**. Open the component menu and change the value of the power source to 50 V (use the **Value** tab). Go to the Display tab and activate Show reference ID and Show values, click on OK, and the designations V1 and 50 V will now be displayed.

6. The **Ground** (common) symbol, displayed in Figure 2-3, is a connection point in most electronic circuits and represents a common point of voltage reference for a circuit. This common point is used for voltage measurements as a point of reference. When constructing a MultiSIM circuit, make sure that you always place a ground somewhere in the circuit (particularly in AC circuits) to ensure that the test instruments and the circuit will work properly.

Figure 2-3   The MultiSIM Ground Symbol

7. **Inductors** consist of coils of wire and store electrical energy in an electromagnetic field. Inductors may also be called **coils**. In MultiSIM, inductors may have a component designator, an inductance value, and special labels whenever they are needed. In Figure 2-4, the component designators (L1, L2, L3, and L4) and individual inductance values are shown near the components. L1 and L2 are standard inductors (L2 is a virtual inductor). L3 and L4 are variable inductors (L4 is a virtual inductor).

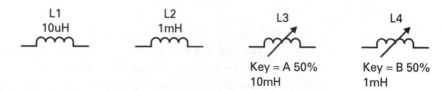

Figure 2-4 MultiSIM Inductor Symbols

8. Open circuit file **02-01c**. Open the component menu for each coil and change the value (Value tab) of the virtual inductors to 100 mH. Go to the **Display** tab for each component, activate **Show reference ID**, and click on **OK**. The component designators L1 and L2 and related inductance values will be displayed.

9. **Capacitors** consist of two conducting plates separated by an insulator. They store electrical energy in an electrostatic field. In MultiSIM, capacitors may have a component designator, a capacitance value, and special labels when needed. In Figure 2-5, the component designators (C1, C2, C3, and C4) and the capacitance values are shown near the component symbols. C1 and C2 are standard capacitors (C2 is a virtual capacitor). C3 and C4 are variable capacitors (C4 is a virtual capacitor).

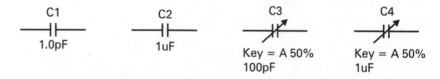

Figure 2-5 MultiSIM Capacitor Symbols

10. Open circuit file **02-01d**. Open the component menu and change the value (Value tab) of both capacitors to 10 uF. Go to the **Display** tab and activate **Show reference ID** and **Show values**, click on **OK** and the component designations (C1 and C2) and related capacitance values will now be displayed.

11. **Transformers** consist of two or more coils of wires (inductors) whose electromagnetic fields have an interacting relationship due to the proximity of their windings. In MultiSIM, a transformer may have a component designator and a special label. In Figure 2-6, the component designators and additional information concerning transformer types are shown above the components. The types of transformers displayed in Figure 2-6 on page 14 are Audio (T1), Power (T2), Radio Frequency (RF – T3) and Virtual (T4).

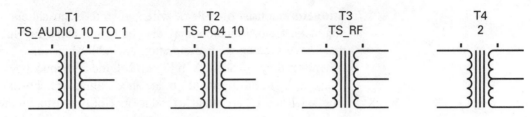

Figure 2-6 MultiSIM Transformer Symbols

12. Open circuit file **02-01e**. Go to the **Display** tab and activate **Show reference ID**, click on **OK** and the designation **T1** will now be displayed. The value of this transformer cannot be changed unless the Edit menu is accessed, which is much beyond the capability of beginning students. The turns ratio of the virtual transformer can be changed quite easily with the menu, something that will be attempted later.

13. **Switche**s are components that open or close electrical circuits and provide a path for current flow when activated. In MultiSIM, switches have a component designator and special labels when needed. In Figure 2-7, the component designator (J1) is shown above the first switch symbol, a single-pole, double-throw switch. The designator J2 is shown above the second switch, a single-pole, single-throw switch. In MultiSIM, a switch is more than a symbol; it will actuate a virtual circuit. That is the purpose of the [Space] designation above the symbol indicating that the **Spacebar** on the computer keyboard will virtually activate the circuit. It is possible to change the actuating key from [Space] to other keys selected from the Value tab drop-down menu. For example, if there are a number of switches on the workspace, each one of them can be actuated by a different key.

Figure 2-7 MultiSIM Switch Symbols

14. Open circuit file **02-01f**. Notice that both switches have the same value designator [Space]. Toggle the [Space] bar on the keyboard and observe that both switches toggle, one up and the other down. Open up the **Component Properties** menu, go to the **Display** tab for each switch and activate **Show reference ID**, click on **OK** and the left switch is designated as "J1" and the right switch is designated as "J2." Now open the menu for Switch J2, open the **Value** tab, and change the designation [Space] to A. Return to the workspace and notice that the A key on the keyboard now toggles Switch J2.

15. **Fuses** are used to protect electronic circuits and are located in the MISC (miscellaneous) library on the left tool bar. In Figure 2-8, the component

designator (F1) and the fuse value (1 A) are shown near the fuse symbol. The fuse is a resistive device (with a different symbol from a resistor) and is identified as a fuse by the F designator.

F1

1_AMP

Figure 2-8 The MultiSIM Fuse Symbol

16. Open circuit file **02-01g**. Notice that the fuse is installed in the circuit with a value of 5 A and is designated as "F1." Activate the circuit by toggling the **Run/stop** simulation switch on the upper toolbar (there are a "0" and a "1" on the switch) or use the **F5** function key to actuate the circuit. Turn the circuit on with Switch J1 (spacebar) and observe that the meter reads 5.0 A.

17. Open up the fuse menu (Value tab) and change the fuse value to 1 A. Reactivate the circuit and observe that the fuse "blows."

18. Return the value of the fuse back to 5 A, reactivate the circuit again, and verify that the circuit has returned to normal operation.

19. **Indicators** are used to indicate on/off conditions in equipment. They can also be used for illumination. They are located in the **Indicator** library on the left toolbar. Figure 2-9 displays a typical lamp indicator. Notice the voltage and wattage ratings—very important data when using lamps.

X1

12V_10W

Figure 2-9 An Indicator

20. There are many other component symbols used in MultiSIM. We will discuss these additional symbols and use them as we progress through the textbook and as it becomes necessary to use these additional symbols in more advanced circuits.

Activity 2.2: The Basic Circuit

1. The minimum requirements for a basic circuit are (1) a voltage source, (2) a load (in this case a lamp, which is a resistive device), (3) a control device (such as a switch), and (4) the method of connection (such as wires or a

printed circuit board). Figure 2-10 shows all of the necessary components that constitute a basic circuit.

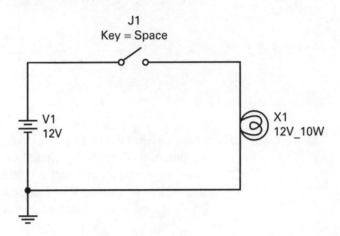

Figure 2-10   A Basic Circuit

2. Open circuit file **02-02**. In this file a basic circuit constructed on the workspace in MultiSIM is displayed.

3. There is an ammeter located on the workspace below the circuit. Insert the meter into the circuit by dragging the meter onto the wire between TPB and TPC. The meter should connect itself and will then indicate the value of current flowing in the circuit.

4. Activate the circuit and close Switch J1 so that current flows in the circuit. What is the value of voltage applied (V_A) to the circuit? Displayed V_A = _____.

5. The term V_A designates voltage applied to a circuit.

6. What is the amount of current (I_T) flowing in the circuit as indicated by meter U1?

   Displayed I_T = _____.

7. I_T designates total current flowing in the circuit.

Activity 2.3: Changing and Measuring Resistance Values

1. When you change the value of a resistor in a circuit there is a corresponding change in the amount of current flowing in the circuit.

2. Open circuit file **02-03a**. Activate the circuit and determine the value of V_A. Sometimes the value of V_A is displayed near the voltage source and other times you have to determine V_A by the use of circuit parameters or a test

instrument. These are the various means necessary to determine the value of the voltage in the circuit being tested. In this circuit, the voltage is indicated to the right of the voltage source (V_A) and by the voltmeter to the left of the voltage source.

Measured $V_A =$ _____.

3. What is the value of I_T displayed by U_1?

   Measured $I_T =$ _____.

4. Change the value of R_1 to 2 Ω. **R1** designates resistor number 1. Activate the circuit. What is the new value of I_T? The new value of $I_T =$ _____.

5. Predict the value of I_T if R_1 was changed to 6 Ω.

   Predicted $I_T =$ _____.

6. Change the value of R_1 to 6 Ω and verify the prediction. Did they match?

   Yes_____ or No_____.

7. Open circuit file **02-03b**. In this activity the multimeter (DMM) will be used to measure resistance. The DMM settings to be highlighted are the "Ω" symbol and the straight line (which represents DC). Move the red probe presently connected to R_1 to R_2 and then to R_3, and so on, in turn, and enter the resistance value for each resistor in Table 2-1. The value of resistor R_1 is already entered as an example of how to enter the data.

	Measured Value		Measured Value
R_1	15kΩ	R_6	
R_2		R_7	
R_3		R_8	
R_4		R_9	
R_5		R_{10}	

Table 2-1 Measuring Resistance

● *Troubleshooting Problems:*

8. Open circuit file **02-03c**. In this exercise, you will measure the resistance of a group of resistors. In the group, three of the resistors will have a measurement that disagrees with the schematic values. Start by entering the schematic values into Table 2-2, measure the resistance of each component, and enter that data into the table. Locate the "bad parts" using the obvious

discrepancies in measured resistance values from the correct values entered into the table.

	Schematic Value	Measured Value
R_1		
R_2		
R_3		
R_4		
R_5		
R_6		
R_7		
R_8		
R_9		
R_{10}		

Table 2-2 Locating Defective Resistors

9. Which three components have an incorrect value? R_____, R_____

    and R_____ are the three resistors that measure incorrectly according to the stated schematic values. The rest agree with the schematic and are correct.

Activity 2.4: Changing Voltage Values and Related Circuit Changes

1. In Activity 2.2, Steps 1 through 5, you learned that by changing resistance in a series circuit, there was a corresponding change in the value of current flow. The next step is to investigate what happens when the value of voltage applied to a circuit is changed.

2. Open circuit file **02-04**. Activate the circuit. What are the values of V_A and I_T?

    Displayed V_A = _____. Measured I_T = _____.

3. Change the value of V_A to 10 V. Activate the circuit and determine the new value of I_T? Measured I_T = _____.

4. Predict the value of I_T if V_A is changed to 5 V. Predicted I_T = _____.

5. Now change the value of V_A to 5 V and verify the prediction. Did they match?

    Measured I_T = _____.

Activity 2.5: Resistor Color Codes (Three- and Four-Band Resistors)

1. Historically, resistors have been found in three-band (20% tolerance) and four-band color code configurations.

2. Open circuit file **02-05**. Activate the circuit and notice the resistance values of the ten resistors according to the displayed schematic.

3. The ten resistors shown are three-band and four-band resistors. Determine the color code according to the schematic values for the resistors and enter the answers in Table 2-3. Remember that 20% resistors do not have a fourth color band.

	Color Band 1	Color Band 2	Color Band 3	Color Band 4
R_1				
R_2				
R_3				
R_4				
R_5				
R_6				
R_7				
R_8				
R_9				
R_{10}				

Table 2-3 Three- and Four-Band Resistor Color Codes

Activity 2.6: Resistor Color Codes (Five-Band Resistors)

1. Color-coded resistors are also found in several five-band configurations: one type with the fifth band indicating tolerance (precision resistors) and another type with the fifth band indicating reliability (specified reliability factor resistors). In this study, five-band **precision** resistors will be studied.

2. Open circuit file **02-06**. The resistors shown are five-band resistors. Determine the proper color code for the ten resistors and enter the resistance values into Table 2-4 on page 20.

	Color Band 1	Color Band 2	Color Band 3	Color Band 4	Color Band 5
R_1					
R_2					
R_3					
R_4					
R_5					
R_6					
R_7					
R_8					
R_9					
R_{10}					

Table 2-4   Five-Band Resistors

Activity 2.7: Adjusting and Measuring Potentiometer Resistance

1. Potentiometers are variable resistors and, when they are used in MultiSIM, they can be varied by the keyboard. Open circuit file **02-07a**. The setting of a potentiometer (pot) can be changed by depressing the key on the keyboard that is related to that potentiometer. For example, the resistance of R_1 in the circuit can be decreased by depressing the key "**R**" and increased by depressing the key "**r**." R_2 uses "**S** and **s**," R_3 uses "**T** and **t**," and so on.

2. To determine total resistance at a particular setting of a potentiometer in MultiSIM, multiply the total resistance of the potentiometer by the percentage displayed to the right of the potentiometer. For instance, the potentiometers displayed in Figure 2-11 are all adjusted to the 50% point in total wiper travel, and their total resistance from the bottom terminal to the wiper would be one-half (50%) of their total resistance. In other words, the percentage setting refers to the resistance from the bottom terminal to the wiper. Of course, total potentiometer resistance is measured from the top terminal to the bottom terminal. The default setting for "change of resistance" is in 5%

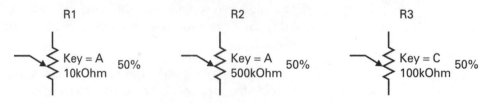

Figure 2-11   Potentiometers in MultiSIM

increments. To change this, double left-click on the potentiometer to bring up the component menu and change the **Increment** setting to the desired setting (for example, 1%). Notice that, in this menu, you can also change the keyboard key that affects the potentiometer. With virtual potentiometers, the resistance value of the potentiometer can be changed. With the non-virtual potentiometers, the resistance value has to be selected from a menu. Also, with non-virtual potentiometers (in the current version of MultiSIM), you will have to type in the potentiometer value under the **Label** tab.

3. Use the DMM to measure total resistance of each potentiometer and enter the data in Table 2-5. Notice that the DMM is connected and ready to measure the total (top to bottom) resistance of potentiometer R_1. Toggle between lower case (r, for example) and upper case (R, for example) to increase and decrease the potentiometer settings.

4. Measure the resistance of each potentiometer from the bottom terminal to the wiper and enter the results in the "Resistance at Starting Setting" column in Table 2-5. Then adjust each potentiometer with the proper keyboard key for readings at both the 25% point and the 75% point. Record the results in Table 2-5.

	Total Resistance of Potentiometer	Resistance at Starting Setting	Resistance at 25% Setting	Resistance at 75% Setting
R_1 [R]				
R_2 [S]				
R_3 [T]				
R_4 [U]				
R_5 [V]				

Table 2-5 Varying Potentiometer Resistance

5. Next, open circuit file **02-07b**. Activate the circuit and record the ammeter and voltmeter readings, with R_1 set for 100%, in Table 2-6.

	Ammeter Reading(I_T)	Voltmeter Reading (V_{R1})
R_1 at 100%		
R_1 at 75%		
R_1 at 50%		
R_1 at 25%		

Table 2-6 Varying Voltage and Current with a Potentiometer

6. Change the potentiometer (R_1) settings to 75%, 50%, and 25%. Record the total current in the circuit (Ammeter – I_T) and the voltage drop across R_1 (Voltmeter – V_{R1}) for each setting of the potentiometer. This project demonstrates that the voltage drop across the potentiometer is proportional to the portion of total circuit resistance that the potentiometer represents. For example, if the potentiometer is 500 Ω at a 50% setting, then it only represents one-third of the total circuit resistance (1500 Ω) and will only drop one-third of the applied voltage.

Activity 2.8: Validating Ohm's Law

1. Georg Simon Ohm formulated the relationship between voltage, current, and resistance in 1827. His contemporaries laughed at his newly-stated formula, so much so that Ohm resigned his professorship in Germany and departed. Soon, the validity of Ohm's formula (law) was found to be accurate and Ohm returned to his teaching. His law states that the amount of current that flows in an electrical circuit is directly proportional to the amount of voltage applied to the circuit and indirectly proportional to the resistance of the circuit. In other words, the more voltage there is, the more current flow there is in the circuit; and the more resistance there is to oppose the current flow, the less current flow there is in the circuit. The relationship can be shown mathematically as: **$I = V/R$**.

2. The first step is to validate the concept of the "directly proportional" relationship between current and voltage. Open circuit file **02-08a**. The resistance of R_1 is going to remain at 1 kΩ throughout this experiment. Only the voltage will be changed to observe (and record) the resulting change in circuit current flow.

3. Change the value of the voltage applied (V_A) to the circuit from 2 V to 4 V, 6 V, 8 V, 10 V, and 12 V. Record the corresponding change in the value of I_T, in Table 2-7.

	$V_A = 2$ V	$V_A = 4$ V	$V_A = 6$ V	$V_A = 8$ V	$V_A = 10$ V	$V_A = 12$ V
$I_T =$						

Table 2-7   Proportionality of Voltage and Current

4. Notice that the current increased proportionally with the increase in voltage. What would happen if the voltage were changed to 20 V? Calculated $I_T =$

_____. Change the voltage to 20 V and verify your calculation. Do not forget to use scientific notation and unit of measurement, for example,

37 mA. Measured $I_T =$ _____.

5. The second step is to validate the "inversely proportional" relationship between current and resistance. Open circuit file **02-08b**.The voltage of the voltage source V_A is going to remain at 10 V throughout this experiment.

6. Change the value of the resistance (R_1 from 1 kΩ, to 2 kΩ, to 5 kΩ, to 10 kΩ and to 20 kΩ, in turn, recording the changes in I_T with each change in resistance value in Table 2-8.

	$R_1 = 1$ kΩ	$R_1 = 2$ kΩ	$R_1 = 5$ kΩ	$R_1 = 10$ kΩ	$R_1 = 20$ kΩ
$I_T =$					

Table 2-8   Inverse Relationship of Resistance and Current

7. Notice that the current flow decreased proportionally as the resistance increased. Calculate what would happen to current if the amount of resistance in the circuit were changed to 500 Ω? Calculated $I_T =$ _____.

8. Change the value of R_1 to 500 Ω and notice the change in current flow. Measured $I_T =$ _____.

9. Now that the relationships between current, voltage, and resistance are understood, it is possible to calculate any one of the three variables simply by knowing the value of the other two variables.

10. Open circuit file **02-08c**. Notice that a different key on the keyboard activates the on/off switch for each circuit. In Circuit 1, calculate the value of I_T using the known values of V_1 and R_1. Calculated $I_T =$ _____.

11. In Circuit 2, activate the circuit and calculate the value of V_2 using the known values of R_2 and I_{T2}. Calculated $V_2 =$ _____.

12. Connect the DMM across R_2 (at the junctions above and below the resistor) and measure the voltage (V_{R2}). Measured $V_{R2} =$ _____.

13. In Circuit 3, activate the circuit and calculate the value of R_3 using the known values of V_3 and I_{T3}. Calculated $R_3 =$ _____. Use the DMM (XMM1) to measure R_3. Measured $R_3 =$ _____.

● ***Troubleshooting Problems:***

14. Open circuit file **02-08d**. There is something wrong with this circuit. According to the Ohm's law calculations using the voltage and resistance values of the circuit, the displayed value of total current is incorrect. The current should be _____, but I_T displays _____. Use the DMM and

measure V_1 and R_1 to determine the problem(s). The problem(s) is/are _____. (You should have determined that there are two problems in this circuit.)

15. Open circuit file **02-08e**. This circuit has one problem. Calculate current from the displayed values of V_1 and R_1. Calculated $I_T =$ _____.

16. Use the DMM to measure I_T. Measured $I_T =$ _____.

17. What is the problem in this circuit? The problem is

_____.

Activity 2.9: Calculating Power Consumption

1. **Power** can be defined as the rate of using energy, and **energy** is the ability to do work. **Electrical energy** is the ability to do electrical work. The rate of expenditure of that electrical work is measured in a unit called the **watt**. The basic formula for electrical power is "Power (in watts) is equal to voltage (in volts) times current (in amperes)" or $\mathbf{P = V \times I}$.

2. Open circuit file **02-09a**. Calculate how much power the circuit is consuming (Use the $P = V \times I$ formula). Calculated $P_T =$ _____.

3. By combining the power formula with the Ohm's law formula, other mathematical methods of determining power consumption are possible. The three formulas are: $\mathbf{P = V \times I}$, $\mathbf{P = V^2 / R}$, and $\mathbf{P = I^2 \times R}$.

4. Open circuit file **02-09b**. How much power is the circuit consuming? $P_T =$ _____. (Use the $P = V^2 / R$ formula).

5. Open circuit file **02-09c**. How much power is the circuit consuming? (Use the $P = I^2 \times R$ formula). Calculated $P_T =$ _____.

● *Troubleshooting Problem:*

6. Open circuit file **02-09d**. There is something wrong with this circuit. The circuit should consume 720 W of power according to the stated voltage and the resistance values. The circuit is only consuming 360 W of power. Use the DMM to measure V_1 and R_1 and, thereby, determine the problem. The

problem is _____.

3. Electric Circuits

References
Electronics Workbench®, MultiSIM Version 7
Electronics Workbench®, MultiSIM Version 7 User's Guide

Objectives After completing this chapter the student should be able to:
- Determine the requirements for a complete circuit
- Recognize a series circuit configuration
- Recognize a parallel circuit configuration
- Recognize a series-parallel circuit configuration
- Determine current paths in a complex circuit
- Measure voltage and current in a circuit
- Troubleshoot a circuit using voltage and current meters

Introduction

One of the job requirements for electrical/electronic technicians is to be able to troubleshoot electronics equipment by comparing the electrical parameters of a circuit under test with the schematic diagram of the circuit. The technician has to be able to interpret the diagram and break the circuit down into discrete segments that are recognizable as series, parallel, series-parallel, and even more complex combinations of components. The schematics of electrical circuits are usually much more complex than the circuits that are studied in the class-room and constructed in the school laboratory. During the troubleshooting process, it is necessary to break these complex circuits down into basic building blocks, and then to isolate the problem. The building block concept is used in electronics troubleshooting by means of the block diagram, a type of drawing. These blocks are the various combinations of components that make the circuit work.

Knowledge concerning electrical circuits in general and the ability to properly use electronics test equipment to determine the operation of a circuit under test is a normal job requirement for all electrical/electronic technicians. Using the voltmeter, ammeter, ohmmeter, and other pieces of test equipment to troubleshoot can be easily learned in the school setting. The student can make use of MultiSIM to build a virtual circuit on the computer monitor and then compare the virtual information with a "real" circuit built on the test bench in the laboratory.

Activity 3.1: Recognizing a Complete Circuit

1. As has been previously stated, the basic circuit consists of a voltage/current source, a load, a control device, and the connecting means (wires, printed circuit boards, and so on). In the circuit of Figure 3-1, the voltage source is a battery power supply (V_A), the load is the lamp (X_1), the control device is the switch (J_1), and the connectors are the wires between the components. On many electronics schematics, the designators **L** for lamp and **S** for switch are commonly used.

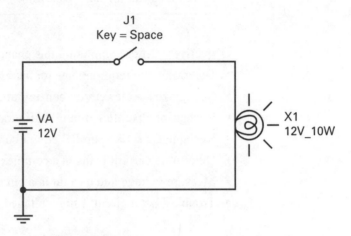

Figure 3-1   A Basic Circuit in MultiSIM

2. Open circuit file **03-01a**. This circuit contains all of the components that are necessary to meet the requirements of the basic circuit. In addition, there is an ammeter (panel type) installed in the circuit to indicate total circuit current flow. Activate the circuit and actuate Switch J1 so that current flows in the circuit. What is the total amount of current (I_T) flowing in the circuit as indicated by ammeter U_1? Measured I_T = _____.

3. Disconnect the wire between TPB and TPC. What is the amount of current (I_T) flowing in the circuit now? Measured I_T = _____.

4. When you see current readings in MultiSIM with the micro (μ) or nano (n) symbol as part of the value indicated on the meter, you should usually interpret these readings as zero readings or very close to that value. This is especially true when a much larger value is expected. This statement has to be taken with a degree of caution because some readings might normally be in the micro/nanovolt or micro/nanoampere region and the μ/n symbol is to be expected in those cases. You, as a technician, should always have a rough approximation of what quantity of voltage or current to expect before making a measurement.

5. With the connection removed between TPB and TPC, is the circuit complete? Yes_____ or No_____. Restore the connection between TPB and TPC.

6. Open circuit file **03-01b**. In this circuit, the meter (**XMM1**) is prepared to measure current. Activate the circuit. Double left-click on the meter (to be able to read it) and close switch J_1. What is the current reading? Notice the meter settings and the probe connections; you will be manipulating the meter switches and making the connections in future exercises. Measured I_T using

   XMM1 = _____.

7. Is the reading the same as in Step 2? Yes_____ or No_____. It should be the same.

8. Open circuit file **03-01c**. In this circuit the meter (**XMM2**) is prepared to measure current; activate the circuit. Double left-click on the meter, turn on the power switch on the meter face, and close switch J_1. What is the current reading? Notice the meter settings and the probe connections. Measured I_T

   using XMM2 = _____.

9. Is the reading the same as Step 2? Yes_____ or No_____. It should be the same.

10. Remove XMM2 from the circuit and install XMM3 in its place. This meter is not set up to read DC current. For the meter to be able to measure current (when it is not preset for you) you will need to turn the instrument on with the **Power** switch, push the **Shift** switch (lower right), and then push the **DC V**

    switch. Use XMM1 and measure the current again Measured I_T = _____.

11. Is the reading the same as in Step 2? Yes_____ or No_____. It should be the same.

● *Troubleshooting Problem:*

12. Open circuit file **03-01d**. Is this a complete circuit? Yes_____ or

    No_____. If your answer is no, take a good look at the circuit to determine the problem? The problem is _____

    _____.

Activity 3.2: Recognizing a Series Circuit

1. A series circuit can be recognized as a circuit that has **one** complete path for current flow; in other words, the current is the same at all points in the circuit. Another attribute of a series circuit is that the sum of the voltage drops across the individual components is equal to the voltage applied to the circuit by the voltage source (V_A).

2. As stated, the current is the same at every point in a series circuit. Open circuit file **03-02a**. This is a series circuit with one path for current flow. This

circuit is similar to previous circuits except that a resistor (**R1**) has been added in addition to the lamp load. This will change the current reading.

3. Activate the circuit and close Switch J1 to complete the circuit. Does Lamp X_1 indicate that it is on? Yes_____ or No_____. Determine the value of I_T as indicated by ammeter U_1? Measured I_T = _____. Has the current value changed from Step 2 in Activity 3.1? Yes_____ or No_____.

4. Open circuit file **03-02b**. This circuit has ammeters installed at three points in the circuit. Activate the circuit and determine whether this is a series circuit. Is this a series circuit? Yes_____ or No_____.

5. Record the ammeter readings. Ammeter U_1 reads _____, U_2 reads _____, and U_3 reads _____.

- *Troubleshooting Problem:*

6. Open circuit file **03-02c**. Activate the circuit and observe whether there is current flow in the circuit. There appears to be something wrong; take a good look at the circuit and determine the problem. The problem is

_____.

Activity 3.3: Recognizing a Parallel Circuit

1. A parallel circuit can be recognized as a circuit with more than one path for current to flow through. The current leaves the voltage source, flows through two or more paths, and returns to the voltage source.

2. Open circuit file **03-03a**. This is a parallel circuit with two lamps connected as a parallel load. Notice that there are two paths for current flow, one through Lamp X1 and the other through Lamp X2.

3. Activate the circuit and close Switch J1 to complete the circuit. Do both lamps indicate that they are on? Yes_____ or No_____. Turn the circuit off (again, make sure that the circuit is deactivated before installing the ammeter).

4. Insert the ammeter (U_1) into the circuit and reactivate the circuit. What is the value of I_T as indicated by Ammeter U_1? Measured I_T = _____.

5. Disconnect Lamp X2 from the circuit by opening Switch J2 with the **A** key Does Lamp X_1 still indicate the "On" condition? Yes_____ or No_____. How about X2; is it still on? Yes_____ or No_____. What happened to current as indicated by Ammeter U_1? Measured I_T = _____.

6. Open circuit file **03-03b**. This circuit has ammeters installed in each parallel branch as well as an ammeter for measuring I_T. Activate the circuit, close Switch J1 and determine the ammeter readings. Measured U_1 = _____, U_2 = _____, and I_T (U_3) = _____.

7. Open Switch J2 and observe the changes in current flow. Record the changed readings. Measured U_1 = _____, U_2 = _____, and I_T (U_3) = _____.

- **Troubleshooting Problem:**

8. Open circuit file **03-03c**. Activate this circuit and make an educated guess on what the problem is. The problem appears to be _____ _____.

9. If the problem seems to be a lamp, which one do you think is bad? Lamp _____ seems to be faulty.

10. In circuits of this type, the most common failure is a bad lamp. Replace the lamp that you think is bad with the spare lamp (X3) placed on the workspace. Did the replacement solve the problem? Yes_____ or No_____.

Activity 3.4: Recognizing a Series-Parallel Circuit

1. Series-parallel circuits have two primary criteria that make them series-parallel circuits. First, the circuit has one or more components that are in series with the power source (in other words, all current flows through them) and secondly, there are also components that are in parallel with other components. In the parallel portions of this type of circuit, the current divides between the components of the parallel branches.

2. Open circuit file **03-04a**. This is a simple series-parallel circuit with three lamps. Lamps X1 and X2 are in parallel with each other, and together, they are in series with X3. All three lamps are in series with the power source.

3. Activate the circuit. Close both of the switches to complete the circuit. Do all of the lamps indicate that they are on? Yes_____ No_____. What is the value of I_T as indicated by ammeter U_1? Measured I_T = _____.

4. Disconnect Lamp X2 from the circuit by opening Switch J2. Which lamps are still on? Lamps X_____ and -_____ are still on.

5. What happened to circuit current (I_T)as indicated by ammeter U_1? Measured I_T = _____.

6. Open circuit file **03-04b**. In this series-parallel circuit there are current meters in each parallel branch as well as several in the series path. Which pairs of meters have the same current indication, and in which portion of the circuit are they installed (series or parallel)? Meters U_1 and _____ have the same reading and both of them are in a **series/parallel** (circle the correct answer) section of the circuit. Meters U_3 and _____ have the same reading, and both of them are in a **series/parallel** (circle the correct answer) section of the circuit.

● **Troubleshooting Problem:**

7. Open circuit file **03-04c**. This is the same circuit as in Step 6 and one of the resistors is open (no current flowing through it). Which resistor seems to be open? Resistor _____ seems to be open.

Activity 3.5: Measuring Voltage in a Circuit

1. Voltage is measured in a circuit by connecting the test leads of the voltmeter across the component or at test points in a circuit where you desire information.

2. Open circuit file **03-05a**. In this circuit, ammeter U_1 measures I_T. Activate the circuit and notice that various meters and DMMs are already connected across the resistors, R_1 through R_4. What is the voltage drop across the resistors? Measured $V_{R1} =$ _____, measured $V_{R2} =$ _____, measured $V_{R3} =$ _____, and measured $V_{R4} =$ _____.

3. Notice that the test leads from the meters are connected so that the positive terminal is connected to the most positive end of the resistor (concerning voltage potential) and the negative terminal is connected to the most negative end. The negative end of any component is always closest in voltage potential to the negative terminal of the power source (marked in this circuit by the negative (−) symbol at the bottom of the power source). If the test leads are reversed when they are connected to the circuit, the display will indicate a negative answer.

4. Open circuit file **03-05b**. Use the DMM to determine voltage drops across each of the resistors in this circuit. Record the data in Table 3-1.

	V_{R1}	V_{R2}	V_{R3}	V_{R4}
Voltage Readings				

TABLE 3-1 Circuit Voltage Drop Data

● *Troubleshooting Problems:*

5. Open circuit file **03-05c**. In this circuit there are two open components. Activate the circuit and notice the current flow. The fact that there is some current flow indicates that the open components are parallel components and not series components.

6. When resistors are in parallel and, seemingly, one of them is open, it is difficult to determine which one is open without visual observation (burnt?) or by disconnecting them one at a time and observing the resultant action on the voltage and current meters. You could also insert a current meter in each parallel branch to check for proper current flow in that branch. In this circuit, when an open resistor is disconnected, there will be no change in the voltage or current reading. When a "non-open" resistor is disconnected, there will be definite changes in the circuit operation as observed on the meters.

7. Disconnect resistors one at a time and determine which ones are open? The open resistors are R_____ and R _____.

8. What would happen to this circuit if R_6 burned open? If R_6 burned open _____. What would happen to total current (I_T) in the circuit? I_T would be = _____.

4. Analyzing and Troubleshooting Series Circuits

References

Electronics Workbench®, MultiSIM Version 7

Electronics Workbench®, Multi SIM Version 7 User's Guide

Objectives   After completing this chapter the student should be able to:

- Recognize a series circuit
- Measure and record current and voltage measurements in a series circuit
- Determine voltage drops in a series circuit
- Prove the accuracy of Kirchhoff's voltage law
- Analyze a series circuit to determine total resistance
- Analyze series circuits to determine voltage, resistance, current, and power requirements of all components in the circuit
- Prove the concept of circuit ground (common)
- Recognize voltage readings as being related to circuit ground or as voltage drops
- Analyze series aiding and opposing voltage sources
- Analyze unloaded voltage dividers
- Troubleshoot series circuits

Introduction

As has been previously stated, series circuits are circuits in which the same amount of current flows through every component in the circuit. As a technician, you have to understand the characteristics of a series circuit to be able to interpret schematic diagrams and diagnose circuit faults.

In a series circuit, any open component in the circuit will prevent current flow. If one of the components changes to a lower value or shorts (zero resistance), then the current flow will increase. At the same time, if one of the components increases in resistance, then the current flow will decrease. The ultimate in resistance, an open circuit, causes current flow to completely cease. With these concepts in mind, it can be stated that resistance and current flow in a series circuit are inversely proportional.

When using schematic diagrams, it is necessary to interpret the various voltage readings on the schematic and compare them with the actual voltage found in the circuit. Many schematic diagrams have "expected" circuit voltages printed on the schematic at key locations to help service technicians troubleshoot defective equipment. There are some companies that produce schematics with test information, key voltages, and oscilloscope waveforms to help the technician in the troubleshooting process.

It is imperative that the technician be able to understand and properly use electronics test equipment to determine the operation, correct or faulty, of the circuit under test and determine which component(s) are defective.

Activity 4.1: The Series Circuit

1. In review, a series circuit consists of a voltage source, a load, a control device, and the necessary connections. The components are connected in such a way that all of the circuit current flows through all of the components. Figure 4-1 is a typical series circuit. Try building this circuit using your MultiSIM program.

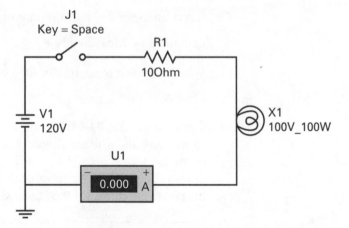

Figure 4-1   A Series Circuit

2. Open circuit file **04-01**. There are three circuits shown; determine which circuit is a series circuit. Circuit _____ is a series circuit. In this circuit, all of the current flows through all of the components. From knowledge gained in the previous chapter, determine the following: Circuit _____ is a parallel circuit and Circuit _____ is a series-parallel circuit.

3. An ammeter (U_1) is connected in the Circuit 2 to measure (I_T). Current flow is always **through** an ammeter, and it has to be connected in series with the rest of the circuit. Activate the circuit and determine I_T. Measured $I_T =$ _____.

4. Use the DMM as a voltmeter to measure the voltage drops in Circuit 2 across the three resistors and record them in Table 4-1.

	V_{R4}	V_{R5}	V_{R6}
Circuit 2 Resistor Voltage Drops			

TABLE 4-1 Voltage Drops in a Series Circuit

Activity 4.2: Measuring Voltage and Current in a Series Circuit

1. Open circuit file **04-02a**. This series circuit has a voltmeter connected across each of the resistors. Activate the circuit and record voltage drops V_{R1}, V_{R2}, and V_{R3} in Table 4-2.

	V_{R1}	V_{R2}	V_{R3}
Resistor Voltage Drops			

TABLE 4-2 More Voltage Drops in a Series Circuit

2. Insert ammeter U_1 into the circuit to measure I_T. Activate the circuit and determine I_T. Measured $I_T = $ _____.

3. What is the voltage output of the voltage source?

   Displayed $V_A = $ _____.

4. Open circuit file **04-02b**. Connect the voltmeters across the three resistors to measure the voltage drops. Activate the circuit and record the voltage drops for V_{R1}, V_{R2}, and V_{R3} in Table 4-3. The positive end of the meter is always connected to the most positive end of the resistor—the end closest to the positive terminal of the power supply.

	V_{R1}	V_{R2}	V_{R3}
Resistor Voltage Drops			

TABLE 4-3 More Voltage Drops in a Series Circuit

5. Insert ammeter U_4 into the circuit to measure I_T. Measured $I_T = $ _____.

6. There is an obvious relationship between the voltage drops across the resistors and the amount of resistance. Using the Ohm's law formula $V = I \times R$, it is possible to calculate the individual voltage drops across the individual resistors. According to Kirchhoff's voltage law for series circuits, the sum of the individual voltage drops is equal to the voltage applied (V_A) to the circuit by the power source.

● *Troubleshooting Problem:*

7. Open resistors prevent current from flowing in a series circuit. An open resistor has infinite resistance, and all of the applied voltage will be dropped across the open. In such a case, the other resistors in such a circuit will measure a 0 V voltage drop (or at least, a very small voltage drop) across each of them. The open resistor(s) in a series circuit is/are the highest resistance(s) (infinite resistance) in the series path, and the other resistors are small in comparison.

8. Open circuit file **04-02c**. Activate the circuit and use the DMM to measure and record the voltage drops across the three resistors. Enter the data for V_{R1}, V_{R2}, and V_{R3} in Table 4-4.

	V_{R1}	V_{R2}	V_{R3}
Resistor Voltage Drops			

TABLE 4-4   More Voltage Drops in a Series Circuit

9. Which resistor is open? R =3 is open. How do you know that R _____ is

   open? R _____ appears to open because _____

   _____.

10. Open circuit file **04-02d**. Calculate I_T for this circuit. Calculated I_T =

   _____.

   Activate the circuit and determine which component(s) is/are faulty. The fault can be an open resistor or power source, a shorted (zero ohms) resistor or power source, resistors that are not the correct value, and missing connections points (in a real circuit, this could be broken wires and connections on a printed circuit board). The faulty component(s) is/are _____

   _____. The resulting fault condition(s) is/are

   _____.

   Use the extra resistors found on the workspace to replace the defective part(s). After replacing the defective component(s) measure circuit current (I_T) using ammeter U_1. Measured I_T = _____.

Activity 4.3: Using Kirchhoff's Voltage Law in a Series Circuit

1. Kirchhoff's voltage law for a series circuit states that "the sum of the voltage drops in a series circuit is equal to the voltage applied to the circuit" or "the sum of the voltage drops is equal to the voltage rise." Each voltage drop has negative to positive polarity from one end of the load to the other. One end of

the load is closer to one power source terminal than the other. There is a negative to positive voltage drop and the end of the load (for example) closest to the negative terminal of the power source is more negative than the other end, which is closer to the positive terminal.

2. Open circuit file **04-03a**. Measure the voltages at test points TPA, TPB, TPC, and TPD in reference to ground (TPE) and record them in Table 4-5. "In reference to ground" means that the negative terminal on the DMM is connected to ground for the purpose of this exercise.

	TPA	**TPB**	**TPC**	**TPD**
Voltage Measurements				

TABLE 4-5 Kirchhoff's Voltage Law in a Series Circuit

3. Notice that the voltages in the circuit, starting at TPD and moving toward TPA, are progressively more positive (in reference to ground – TPE). The negative to positive voltage drops across each load add up to the applied voltage.

4. Measure the voltage at TPE. Voltage at TPE = _____.

- ● **Troubleshooting Problems:**

5. Open circuit file **04-03b**. Activate the circuit and notice that there is a problem; the current flow should be 750 mA and it is too high. What is wrong? Use the DMM to measure resistance and determine the fault. The problem is

    R_____, which _____

    _____.

6. Open circuit file **04-03c**. Calculate I_T and the voltage drops for each of the resistors and enter the voltage drop data in Table 4-6. Calculated I_T =

    _____.

	R$_1$	**R$_2$**	**R$_3$**	**R$_4$**
Calculated Voltage Drops				
Actual Voltage Drops				
Actual Resistance Values				

TABLE 4-6 Determining Voltage Drops in a Series Circuit

7. Activate the circuit and observe I_T. What is the actual value of I_T? Measured

    I_T = _____.

8. Obviously, there is a problem with the circuit current flow as indicated by ammeter U_1. use the Agilent DMM to measure the voltage drops across each of the resistors and enter this data into Table 4-6. Determine whether there is a problem with the voltage drops. If so, what is it? The problem is

   _____.

9. Measure the resistance values. Enter the resistance values in Table 4-6. What is wrong with the circuit? The fault in the circuit is _____

   _____.

Activity 4.4: Determining Total Resistance in a Series Circuit

1. Total resistance in a series circuit is equal to the sum of the individual resistances. As stated previously, you simply add up the individual resistance values.

2. Open circuit file **04-04a**. Calculate total resistance (R_T) in the circuit by adding the individual resistance values. Calculated R_T = _____.

3. Use the DMM to verify your calculation. Measured R_T = _____.

4. Open circuit file **04-04b**. Use the Agilent DMM to determine total resistance (R_T) in this circuit. Measured R_T = _____.

Activity 4.5: Determining Unknown Parameters in Series Circuits

1. If any two factors in a circuit such as voltage, resistance, current, or power consumption are known about a particular component, then the remaining (unknown) factor(s) can be determined about that component. In a series circuit, the most important parameter to know is current flow, because current flow is the same everywhere in the circuit. Use Ohm's law, Kirchhoff's voltage law, and known characteristics about series circuits to solve for the unknown quantities in this activity.

2. Open circuit file **04-05a**. In this circuit, the known parameters for each of the resistors are current flow and voltage drop. Solve for the resistance of the three resistors. Calculated values are: R_1 = _____, R_2 = _____, and R_3 = _____.

3. Open circuit file **04-05b**. Solve for current in this circuit using V_A and R_T. Calculated I_T = _____ μA or _____ mA. Install the Agilent multimeter in the circuit to measure current. Measured I_T = _____.

4. Open circuit file **04-05c**. Solve for voltage in this circuit using Ohm's law, $V_A = I_T \times R_T$. Round off the I_T value for your calculation. Calculated $V_A = $ _____.

5. Use the Agilent multimeter to verify your calculation. Measured $V_A = $ _____.

6. Open circuit file **04-05d**. Solve for voltage in this circuit using Kirchhoff's voltage law, $V_A = V_1 + V_2 + V_3$. Calculated $V_A = $ _____.

7. Open circuit file **04-05e**. Activate the circuit and solve for unknowns, filling in the empty blanks in the following partial solution matrix (Table 4-7). The known factors about the circuit are already entered in the table.

Component	Resistance	Voltage	Current
R_1		4.4 V	
R_2	4.7 kΩ		
R_3		13.6 V	
R_4	10 kΩ		
R_5		6.6 V	
Totals			

TABLE 4-7   Determining Series Circuit Parameters

8. Calculate the power requirements for each of the resistors in Table 4-7. $P_{R1} = $ _____, $P_{R2} = $ _____, $P_{R3} = $ _____, $P_{R4} = $ _____, and $P_{R5} = $ _____.

9. What is the total power consumed by the circuit? $P_T = $ _____.

● *Troubleshooting Problems:*

10. Open circuit file **04-05f**. Using the solution matrix of Step 7 for comparison and the DMM, find the faulty resistor in this circuit and describe the fault. The faulty resistor is R_____ because _____ _____.

11. Open circuit file **04-05g**. Using the solution matrix of Step 7 for comparison and the Agilent DMM, find the faulty resistor in this circuit and describe the fault. The faulty resistor is R_____ because _____ _____.

12. Open circuit file **04-05h**. Using the solution matrix of Step 7 for comparison and the DMM, find the faulty component in this circuit and describe the fault. The faulty component is _____ because _____

    _____.

13. Open circuit file **04-05i**. Using the solution matrix of Step 7 for comparison and the Agilent® DMM, find the faulty component in this circuit and describe the fault. The faulty component is _____ because _____

    _____.

Activity 4.6: Grounding the Circuit

1. In all circuits there is a point of reference that is considered to be 0 V. This point is referred to as **circuit ground** or **circuit common**. Usually this point in the circuit is the negative terminal of the power source, but it could also be the positive terminal. Ground, in a circuit, is wherever the designer determines that a common or reference connection point needs to be located. This is the point of reference when voltages are being measured or compared. This excludes voltage drop measurements when a voltage measurement is made across a component or between two points in the circuit rather than from ground. When measuring a voltage, decide whether the measurement is in reference to ground, which requires the connection of the negative terminal of the DMM to the ground, or if a voltage drop measurement is being made where the test leads are connected across the device being measured.

2. Open circuit file **04-06a**. Ground is connected to the negative terminal of the power source (normal connection). Measure the voltage readings at the four test points and record the data on the first line (ground at TPD) in Table 4-8.

Ground Locations	Voltage at Test Points in Reference to Ground			
	TPA	TPB	TPC	TPD
At TPD				
At TPC				
At TPB				
At TPA				

TABLE 4-8 Varying Ground Locations

3. Move the ground terminal from TPD to TPC. Measure the voltages again and record the data on the second line (ground at TPC) of Table 4-8.

4. Move the ground terminal from TPC to TPB. Measure the voltages again and record the data on the third line (ground at TPB) of Table 4-8.

5. Move the ground terminal from TPB to TPA. Measure the voltages again and record the data on the fourth line (ground at TPA) of Table 4-8.

6. Notice that some of the voltages become negative when the ground location is moved away from the negative terminal of the power source toward the positive terminal. In fact, with the ground at TPA, all of the readings are negative in respect to ground. Return the ground terminal connection to TPD where it was originally located.

7. When referring to voltage drops such as the voltage drop across R2 (TPB to TPC), the voltage would be referred to as V_{R2} or V_{BC}. The first reference letter of V_{BC} (the B) indicates the measurement point and the second letter (the C) indicates the reference point. Measure voltage V_{AC}. Measured V_{AC} = _____.

8. Usually, V_A indicates the voltage applied to the circuit and not the voltage at TPA in reference to ground. If there were a TPA in the circuit, then V_A would be the same as TPA. What is the voltage at TPC (V_C)? Measured V_C = _____.

● **Troubleshooting Problems:**

9. Open circuit file **04-06b**. Activate the circuit and close the switch. Smoke is rolling out of the power source. What is wrong? The problem is _____.

10. Normally, a fuse would protect a power source of this type, or there would be internal circuitry that would prevent a massive overload such as this. Notice that the current being drawn from the power source is 12 GA according to ammeter U_1? How many amps are 12 GA? 12 GA is _____ Amperes.

11. Open circuit file **04-06c**. Activate the circuit and determine the problem. There is no current flow in the circuit according to the ammeter. But, the voltmeter is measuring voltage drop, which indicates that current is flowing through R_1. The problem is _____.

Activity 4.7: Series-Aiding and Series-Opposing Power Sources

1. In electronics equipment, such as personal computers, many different power sources are needed to operate all of the various types of circuits with their differing voltage and current requirements. Sometimes the needed voltages are positive and sometimes negative. In some circuit configurations the power

sources aid one another and in other circuit configurations the power sources oppose one another.

2. When power sources aid one another, the output voltages add. Open circuit file **04-07a**. What do the voltages being displayed by U_1, U_2, and U_3 represent? U_1 displays the V_1 voltage of 5 V, U_2 displays _____, and U_3 displays _____.

3. When power sources oppose one another, the output voltages algebraically add. For example, in a two-cell flashlight, if you place one of the batteries in a reverse situation from the other battery, the two batteries oppose each other and the lamp won't light.

4. Open circuit file **04-07b**. What are the values of the three voltages displayed on U_1, U_2, and U_3? U_1 displays _____, U_2 displays _____, and U_3 displays _____.

5. Open circuit file **04-07c**. What is the algebraic sum of the four voltage sources, (V_{AE}) as measured between TPA and TPE, in reference to ground? Use the Agilent DMM to make your measurements. Calculate V_{AE}. Measured V_{EA} = _____.

6. Use the DMM to determine the value of V_A, V_B, V_C, V_D, and V_E in reference to ground. V_A = _____, V_B = _____, V_C = _____, V_D = _____, and V_E = _____.

7. Calculate the amount of current flowing through R_1? Calculated I_{R1} = _____. Verify the calculation with the DMM. Measured I_{R1} = _____.

- **Troubleshooting Problem:**

8. Open circuit file **04-07d**. The circuit is designed so that the voltage drop across R_1, as indicated by voltmeter U_1 should be 60 V. There is an obvious problem. What is the problem? The defective reading is caused by

_____.

Activity 4.8: Unloaded Voltage Dividers and Resistor Power Requirements

1. A voltage divider is a series of resistors that are connected in an arrangement that provides specific voltages at junctions between the resistors. A voltage divider is analyzed by determining the voltages at the junctions between the resistors in reference to ground.

2. Open circuit file **04-08a**. Measure the voltages at TPA, TPB, and TPC. Measured voltages at TPA = _____, TPB = _____, and TPC = _____.

3. Use the DMM to determine how much current is flowing in the circuit? Measured I_T = _____.

4. Use I_T and the voltage drops across each resistor to determine the power requirements of the resistors in this circuit? Calculated P_{R1} = _____, P_{R2} = _____, P_{R3} = _____, and P_{R4} = _____.

5. What is the total power consumed by the circuit? P_T is the sum of the individual power requirements for all of the resistors in the circuit. Calculated P_T = _____ W.

● *Troubleshooting Problems:*

6. Open circuit file **04-08b**. This circuit is the same as the circuit of Steps 2 through 4. Use the data gathered in those steps (I_T and resistor power requirements) for this problem. Enter the previously determined resistor power requirements into the first row of Table 4-9.

	R_1	R_2	R_3	R_4
Correct Wattage Rating				
Current Wattage Setting				

Table 4-9 Applying Resistor Wattage Ratings

7. Activate the circuit and observe the action of two resistors (R_1 and R_4) as they (virtually) smoke and burn up. Use the component menu for each resistor to ascertain the actual wattage ratings (settings) and enter this data in the second row of Table 4-9.

8. Change the wattage ratings for R_1 and R_4 to reflect your calculations. Activate the circuit and verify the settings. Did the corrected circuit operate properly?

    The circuit operated properly. Yes_____ No_____.

5. Analyzing and Troubleshooting Parallel Circuits

References

Electronics Workbench®, MultiSIM Version 7

Electronics Workbench®, MultiSIM Version 7 User's Guide

Objectives After completing this chapter the student should be able to:

- Recognize a parallel circuit
- Measure and record current and voltage measurements in a parallel circuit
- Determine current paths in a parallel circuit
- Prove the accuracy of Kirchhoff's current law for parallel circuits
- Analyze a parallel circuit to determine total resistance
- Analyze a parallel circuit to determine voltage, current, resistance, and power requirements of all components in the circuit
- Troubleshoot parallel circuits

Introduction

Parallel circuits are circuits in which the voltage is the same across all of the parallel components and in which current divides between the parallel branches. Parallel circuits are commonly found in residential wiring where most of the AC loads (outlets) in a residence are in parallel. Another example is in automotive circuits where all of the electrical devices are in parallel across the DC power source. It is necessary for technicians to understand the characteristics of parallel circuits to enable them to interpret schematic diagrams and diagnose circuit faults in these types of circuits.

In a parallel circuit, a component can be open in one parallel branch and the current will still flow through the other parallel branches. If one of the parallel components changes to a lower value of resistance or shorts out (zero resistance), then the current flow will increase through that component alone and may cause an overloading condition for the power source, which could result in a tripped circuit breaker or a blown fuse. If one of the parallel components increases in resistance, then the current flow will decrease through that component and not affect the other parallel branches. An open parallel branch will have little effect other than to reduce the total current drain on the power source as load decreases.

Activity 5.1: The Parallel Circuit

1. In a manner similar to series circuits, a parallel circuit consists of a voltage source, two or more loads in parallel, a control device, and the necessary connections. All parallel components are connected in such a way that the source current divides between the parallel branches. A parallel circuit can be thought of as being similar to a river in which the current divides to go around an island and then joins again after passing the island. Figure 5-1 displays a typical parallel circuit.

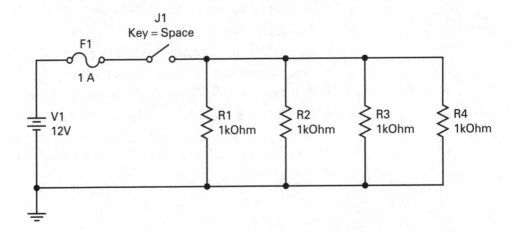

Figure 5-1   A Typical Parallel Circuit

2. Open circuit file **05-01**. How many parallel branches are there in the circuit?

    There are _____ branches in the circuit.

3. In this circuit, the meters are connected in a manner to demonstrate the path of current flow, leaving the power source, dividing between the branches (going around the islands), and then rejoining before proceeding on to the other power source terminal. The U_1 and U_2 meter readings represent total current, and the U_3, U_4, and U_5 meter readings represent the branch currents.

4. Activate the circuit and enter the current flow data in Table 5-1. You can see that the current flowing into the branches is equal to the current flowing out

from the branches and that total current is equal to the sum of the branch currents. What is the total current I_T as indicated by U_1 and U_2? Measured $I_T =$ _____ .

	I_{R1}	I_{R2}	I_{R3}
Resistor Currents in Each Branch			

Table 5-1   Branch Current Data

5. Place R_4 in parallel with the existing circuit by connecting it to TPA and TPB. What happened to I_T? I_T changed from _____ to _____ . Use the DMM to determine how much current is flowing through R_4? Measured $I_{R4} =$ _____ .

6. Whenever another resistor is placed in parallel with the rest of the circuit, I_T increases/decreases (circle the best answer).

Activity 5.2: Measuring Voltage and Current in a Parallel Circuit

1. The current in each parallel branch is equal to the voltage across that branch divided by the resistance of that branch ($I_{Branch} = V_{Branch} \times R_{Branch}$).

2. Open circuit file **05-02a**. Activate the circuit and measure the voltage output of the voltage source (V_A) with the Agilent® DMM? Measured $V_A =$ _____ .

3. Determine I_T with ammeter U_1. Measured $I_T =$ _____ .

4. Measure the voltage across each of the resistors and enter the data in Table 5-2.

	R_1	R_2	R_3
Calculated Currents in Each Branch			
Measured Voltage Across Each Branch			

Table 5-2   Branch Voltage and Current Calculations and Measurements

5. Calculate and record the current through each resistor in Table 5-2. Does the sum of the branch currents equal total current (I_T)? Yes_____ or No_____.

6. There is an obvious relationship between the current flow in each branch of a parallel circuit and the amount of resistance in each branch. Using Ohm's law, it is possible to calculate the individual currents flowing in each branch of the circuit. Another observation is that the sum of the branch currents in a parallel circuit is equal to the total current of the circuit (I_T).

7. Using Ohm's law ($R_T = V_A/I_T$), determine total resistance R_T. Calculated $R_T =$ _____.

8. Open resistors in a parallel branch prevent current from flowing in that particular branch of the circuit. The total current flowing in the parallel circuit will be less than expected by the amount that should be flowing in the faulty current path (branch). Applied voltage will be dropped across the faulty component, but there will be no current flow. Other branches in the circuit will provide paths for current flow and the total current will be the sum of the conducting paths. When determining current flow for a parallel circuit, and the total current is less than expected (according to the calculations), then one of the branches is likely to be open.

- **Troubleshooting Problems:**

9. Open circuit file **05-02b**. Calculate the expected total current for the circuit. Calculated $I_T =$ _____.

10. Activate the circuit. According to ammeter U_1, I_T is less than it should be. Measured $I_T =$ _____.

11. Use the DMM to measure the current in each branch of the circuit. Notice that the DMM is in position to measure the current in the R_1 branch. Use this as a model and record all branch currents in Table 5-3.

	I_{R1}	I_{R2}	I_{R3}
Branch Currents			

Table 5-3   Branch Currents in a Parallel Circuit

12. Which resistor is open? R _____ is open.

13. What should be the amount of current flowing in the open branch? The current in the open branch should be _____.

14. Open circuit file **05-02c**. Calculate the value of total current.

15. Activate the circuit. What value of current should be flowing through the R_3 branch of the circuit? Calculated I_{R3} = _____.

16. Use the DMM to measure I_{R3}. Measured I_{R3} = _____.

17. What do you think is wrong? The problem is that R_3 has a value of _____ and it should be _____.

18. Open circuit file **05-02d**. In this parallel circuit, the lamp should draw about 833 mA. Calculate how much current should flow through the resistor. What should I_T be in this circuit? Calculated I_{R1} = _____. Calculated I_T = _____. The fuse is sized correctly. Yes_____ or No_____.

19. Calculate the resistance of the lamp and R_T using V_A and I_T for your calculation. Calculated lamp resistance = _____. Calculated R_T = _____.

20. What is causing the fuse to blow? Use the DMM to troubleshoot the circuit. The problem is caused by _____ because _____

_____.

Activity 5.3: Using Kirchhoff's Current Law in a Parallel Circuit

1. Kirchhoff's law states that the algebraic sum of the currents flowing in and out of a junction (node) is equal to zero. In simpler terms, the amount of current flowing into a junction is equal to the amount of current flowing out of the junction. Figure 5-2 displays a parallel with ammeters in all possible

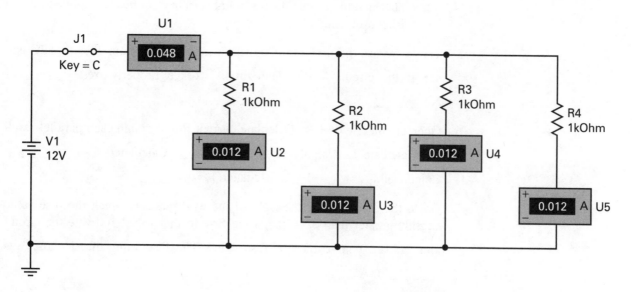

Figure 5-2   Current Paths in a Parallel Circuit

paths of current flow and displaying current into and out of a junction. Notice that

$$I_T = I_{R1} + I_{R2} + I_{R3} + I_{R4}$$

2. Open circuit file **05-03a**. The node equation for this circuit in reference to TPA (Node A) is $I_T = I_{R1} + I_{R2} + I_{R3}$. Calculate the circuit current flow and the branch currents and rewrite the equation using the displayed current values. The node equation for TPA using calculated current values is

_____

_____

3. Activate the circuit. Use the DMM to measure the branch currents and record the data in Table 5-4. Also record I_T.

	I_T	I_{R1}	I_{R2}	I_{R3}
Branch Current Measurements				

Table 5-4   Using Kirchhoff's Current Law

● **Troubleshooting Problems:**

4. Open circuit file **05-03b**. Activate the circuit and notice that there is a problem in this circuit. I_T is too low and there is no current flowing through R_2 and R_3. Use the spare DMM to isolate the problem. The problem is

_____.

Remember that the problem does not always have to be a component; it can be a wiring problem.

5. Try to bypass the problem. Connect a jumper wire from TPA to TPC and see if the circuit works. The circuit worked properly. Yes_____ or No_____.

6. Open circuit file **05-03c**. Calculate current flow through each parallel resistor. Calculate I_T. Calculated I_{R1} = _____. Calculated I_{R2} = _____. Calculated I_{R3} = _____. Calculated I_T = _____.

7. Go to the fuse chart at the right of the workspace and pick the fuse value slightly greater than I_T. Change the fuse to that value. Activate the circuit and validate your choice of fuse value. The fuse value that I chose was _____ mA.

8. Did the circuit operate correctly? Yes_____ or No_____.

Activity 5.4: Determining Total Resistance in a Parallel Circuit

1. Ohm's law is the easiest method to determine total resistance in any circuit and should be used as a first option whenever possible.

2. Open circuit file **05-04a**. Activate the circuit and notice the voltage across the circuit and the current through the circuit. Calculate the total resistance (R_T) of the circuit using Ohm's law. Using Ohm's law, $R_T =$ _____. Open Switch J1 and connect the DMM across the parallel circuit and verify the calculation for R_T. The DMM measures _____.

3. Another method of determining total resistance of a parallel circuit with many branches is by using the **reciprocal method**. This method is valuable when voltage and current parameters are unavailable or difficult to obtain and it is necessary to know the amount of load a circuit offers to a power source. The reciprocal formula is

   $1/R_T = 1/R_1 + 1/R_2 + \ldots + 1/R_n$.

   Simply stated, total resistance in a circuit is equal to the reciprocal of the sum of the reciprocals of all parallel resistance values in the circuit.

   Open circuit file **05-04b**. Calculate the total resistance of Circuit 1 and Circuit 2 using the reciprocal formula. Calculated R_T for Circuit 1 = _____. Calculated R_T for Circuit 2 = _____.

4. Verify the calculations with the DMM. Measured R_T for Circuit 1 = _____. Measured R_T for Circuit 2 = _____.

5. Another method of determining total resistance of a parallel circuit is by using the **conductance method**. Essentially, the conductance method is the same as the reciprocal method because **$G_T = 1/ R_T$**.

6. Open circuit file **05-04c**. Calculate the total resistance of Circuit 1 and Circuit 2 using the conductance formula. Calculated G_T for Circuit 1 = _____. Calculated G_T for Circuit 2 = _____. Calculated R_T for Circuit 1 = _____. Calculated R_T for Circuit 2 = _____.

7. Verify the calculations with the Agilent DMM. Measured R_T for Circuit 1 = _____. Measured R_T for Circuit 2 = _____.

8. Another method of determining total resistance of a parallel circuit is by using the **product-over-sum method, $R_T = (R_1 \times R_2)/(R_1 + R_2)$**. This method works when the circuit consists of only two resistances.

9. Open circuit file **05-04d**. Calculate the total resistance of Circuit 1 and Circuit 2 using the product-over-sum method formula. Calculated R_T for Circuit 1 = _____. Calculated R_T for Circuit 2 = _____.

10. Verify the calculations with the DMM. Measured R_T for Circuit 1 = _____. Measured R_T for Circuit 2 = _____.

11. Another method of determining total resistance of a parallel circuit is by using the **equal-value-resistor method, $R_T = R/N$**.

12. Open circuit file **05-04e**. Calculate the total resistance of Circuit 1 and Circuit 2 using the equal-value-resistor method. Calculated R_T for Circuit 1 = _____. Calculated R_T for Circuit 2 = _____.

13. Verify your calculations with the DMM. Measured R_T for Circuit 1 = _____. Measured R_T for Circuit 2 = _____.

14. In Circuit 2, if another 80-Ω resistor were added in parallel, what would be the new value of R_T? Calculated R_T = _____.

15. Now, place the 80-Ω resistor in the circuit (in parallel with the other resistors) and use the DMM to measure the new value for R_T. Measured R_T = _____.

16. A final method of determining total resistance is by using the **assumed voltage** method. With this method you choose a voltage that is easy to use with all of the resistors. Use this voltage to determine branch currents. Then determine I_T as the sum of the branch currents.

17. Open circuit file **05-04f**. Calculate R_T using the assumed voltage method. What voltage did you choose? I chose _____V. Calculated R_T = _____.

18. Measure R_T with the DMM. Measured R_T = _____.

19. Activate the circuit and adjust the power source to your assumed voltage.

20. What is the value of I_T according to ammeter U_1? Measured I_T = _____.

21. What is the total power (P_T) consumed by this circuit? Use any one of the power formulas to calculate circuit power consumption. Calculated P_T = _____W.

● *Troubleshooting Problems:*

22. Open circuit file **05-04g**. Calculate R_T and then I_T according to the schematic. Calculated R_T = _____ and calculated I_T = _____.

23. What does the ammeter (U_1) indicate I_T is at present? Measured I_T = _____.

24. What is wrong with the circuit? The problem with the circuit is _____

_____.

25. Obviously there are some problems with the resistors. Repair the circuit using the spares located to the right of the circuit. Recheck the circuit. Measured I_T = _____.

26. Open circuit file **05-04h**. When you activate this circuit the fuse blows. Is the fuse the correct value? Yes_____ or No_____.

27. The fuse blows because _____

_____.

Activity 5.5: Determining Unknown Parameters in a Parallel Circuit

1. As previously stated, if any two factors such as voltage, resistance, current, or power consumption are known about a particular component in a circuit, then other unknown factors can be determined about that component. In a parallel circuit, the most important factor to know is voltage, because voltage is the same across all of the parallel components. Use Ohm's law, Kirchhoff's current law, and known characteristics about parallel circuits to solve for unknown factors in this activity.

2. Open circuit file **05-05a**. Solve for the resistance of the three resistors. In this circuit, use branch currents and V_A to calculate the unknown resistors. Calculated R_1 = _____, R_2 = _____, and R_3 = _____.

3. Determine the value of R_T. Calculated R_T = _____.

4. Calculate P_T. What is the power consumption of each of the resistors? Calculated P_T = _____, P_{R1} = _____, P_{R2} = _____, and P_{R3} = _____.

5. Open circuit file **05-05b**. Use Kirchhoff's current law to solve for I_T. Next solve for R_T, and finally find the resistance of each of the resistors. Calculated I_T = _____. Calculated R_T = _____, R_1 = _____, R_2 = _____, and R_3 = _____.

6. Open circuit file **05-05c**. Calculate the voltage drop across each of the resistors. Calculated V_{R1} = _____, V_{R2} = _____, and V_{R3} = _____.

7. Calculate the value of V_A? Calculated V_A = _____.

8. What are the values of I_T and R_T in this circuit? Calculated I_T = _____ and R_T = _____. Verify the calculations with the DMM.

9. Open circuit file **05-05d**. Solve for current in this circuit using Kirchhoff's current law, $I_T = I_{R1} + I_{R2} + I_{R3}$. Calculated $I_T =$ _____.

10. Calculate the value of R_T in this circuit. Calculated $R_T =$ _____.

11. The next circuit uses all of the techniques that you have learned up to this point. You should be able to solve for unknown component values in partially specified parallel circuits.

12. Open circuit file **05-05e**. Measure and calculate to solve for unknowns and fill in the empty blanks in the following partial solution matrix (Table 5-5). Verify the calculations with the DMM.

Component	Resistance	Voltage	Current
R_1			1.667 mA
R_2	12 kΩ		
R_3	21 kΩ		3.571 mA
R_4	15 kΩ		
R_5			7.5 mA
Totals			

Table 5-5   Determining Parallel Circuit Parameters

13. Calculate total power and the power requirements for each of the resistors in Table 5-5. Calculated $P_{R1} =$ _____, $P_{R2} =$ _____, $P_{R3} =$ _____, $P_{R4} =$ _____, $P_{R5} =$ _____, and $P_T =$ _____.

● *Troubleshooting Problems:*

14. Open circuit file **05-05f**. Using the solution matrix of the previous exercise and the Agilent DMM, find the faulty resistor in this circuit and describe its fault. The faulty resistor is R_____ because _____ _____.

15. What does I_T measure? Measured $I_T =$ _____.

16. Is it right? Yes_____ or No_____.

17. What would you do to repair this circuit? I would _____ _____.

18. Open circuit file **05-05g**. Using the results determined in the solution matrix of Step 12 and the DMM, find the faulty resistor in this circuit and describe its fault.

    The faulty resistor is R_____ because _____

    _____.

19. What does I_T measure? Measured I_T = _____.

20. Is the measurement right? Yes_____ or No_____.

21. What would you do to repair this circuit? I would _____

    _____

    _____.

6. Analyzing Series-Parallel Circuits

References

Electronics Workbench®, *MultiSIM* Version 7

Electronics Workbench®, *MultiSIM* Version 7 User's Guide

Objectives After completing this chapter the student should be able to:

- Recognize a series-parallel circuit
- Simplify a series-parallel circuit
- Analyze a series-parallel circuit to determine voltage, current, resistance, and power requirements of all components in the circuit
- Determine characteristics of unloaded and loaded voltage divider circuits
- Observe effects and causes of voltmeter loading
- Recognize and analyze bridge circuits
- Troubleshoot series-parallel circuits

Introduction

Series-parallel circuits are circuits in which the characteristics of series and parallel circuits are combined. It is necessary to use all of the techniques that have been learned up to this point to determine the operational parameters of these more difficult circuits. In electronics equipment, series-parallel circuits are used more often than the other types of circuits that have been studied thus far. When working on equipment it is unusual to find simple series or parallel circuits in the equipment, and it is more common to spend the bulk of the time working on very complex series-parallel combinations. Technicians need to fully understand the characteristics of series-parallel circuits to enable them to interpret schematic diagrams and diagnose circuit faults in electronics equipment.

A series-parallel resistive circuit can be reduced to a single resistance for the purpose of determining total circuit load offered by the circuit. It is possible to construct a particular series-parallel circuit in MultiSIM and determine the operational characteristics of that circuit on the computer monitor without ever touching the actual circuit. In this chapter you will use virtual instrumentation to verify circuit parameters. Figure 6-1 displays a typical series-parallel circuit.

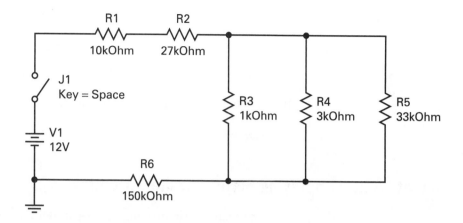

Figure 6-1 A Series-Parallel Circuit

Activity 6.1: The Series-Parallel Circuit

1. As a combination of series and parallel circuits, a series-parallel circuit has the characteristics of both circuits and has to be isolated into discrete sections. Within this type of circuit one portion of the circuit can be a parallel section and another portion can be a series section. It is necessary to take into consideration all of the circuit rules that have been learned up to this point for these two types of circuits.

2. First, determine which components are in series in the circuit. Open circuit file **06-01a**. How many series resistors are there in the circuit? There are _____ series resistors in the circuit.

3. How many resistors are in parallel in the circuit? There are _____ parallel resistors in the circuit.

4. Open circuit file **06-01b**. This has a number of series and parallel sections. Identify them. How many resistors are in series? There are _____ resistors in series in this circuit.

5. How many parallel sections are there in the circuit? There are _____ parallel sections in the circuit.

6. In this circuit, there is one section with three resistors in parallel. The three resistors in parallel are R _____, R _____, and R _____.

7. Calculate the total resistance using the formula $R_T = R_1 + R_2 + (R_3 \parallel R_4) + R_5 + R_6 + R_7 + (R_8 \parallel R_9) + (R_{10} \parallel R_{11}) + R_{12} + R_{13} + (R_{14} \parallel R_{15}) + R_{16} + (R_{17} \parallel R_{18} \parallel R_{19})$. In this text, the $\parallel$ symbol means "in parallel with." Calculated $R_T =$ _____.

8. Use the DMM and measure the total resistance of the circuit. Measured $R_T =$ _____.

9. There are also circuits where resistors are in series within a parallel section. Open circuit file **06-01c**. Calculate the total resistance in this circuit using the formula $\mathbf{R_T = R_1 + (R_2 \parallel [R_3 + R_4]) + R_5 + (R_6 \parallel [R_7 + R_8]) + R_9 +}$ $\mathbf{R_{10} + (R_{11} \parallel R_{12} \parallel [R_{13} + R_{14}]) + R_{15}}$. Calculated $R_T =$ _____.

10. Use the DMM and measure the total resistance of the circuit. Measured $R_T =$ _____.

Activity 6.2: Simplifying Series-Parallel Circuits

1. The secret to simplifying series-parallel circuits is to treat each section as an individual circuit, simplifying the individual sections by reducing them to a single resistance, and then adding up all of the series components along with the reduced (simplified) sections. Generally, the best procedure to follow is to work toward the power source from the point farthest from the power source. In other words, simplify each section and work toward the power source. It is possible to end up with the sum of all the series components and simplified sections becoming a single resistance value that offers a single load to the power source.

2. Open circuit file **06-02a**. This circuit has one parallel section consisting of R_3, R_4, and R_5. To simplify the circuit, calculate the resistance of the parallel section (call the equivalent resistance of that section "R_A"). $R_A =$ _____.

3. To obtain a value for R_T, add the resistance of the reduced section to the three series resistors. Verify your calculation with the Agilent DMM. Calculated $R_T = R_1 + R_2 + R_A + R_6 =$ _____. Measured $R_T =$ _____.

4. Open circuit file **06-02b**. Simplify this circuit. The formula would be $R_T = $ $\mathbf{R_1 + R_2 + [R_3 \parallel R_4 \parallel R_5] + [R_6 \parallel R_7] + [R_8 \parallel R_9]}$. Verify your calculation with the DMM. Calculated $R_T =$ _____. Measured $R_T =$ _____.

Activity 6.3: Measuring Voltage and Current in a Series-Parallel Circuit

1. Open circuit file **06-03a**. Activate the circuit and notice that the current meters demonstrate that current leaves the power source, divides between the branches (going around the islands), and then rejoins before proceeding on to the other power source terminal.

2. Activate the circuit and observe that the DMM and U_1 meter readings represent total current. In this circuit, displayed $I_T =$ _____.

3. The U_2, U_3, and U_4 meter readings represent the branch currents. The total of the branch currents, $I_{R1} =$ _____, $I_{R2} =$ _____, and $I_{R3} =$ _____, is equal to the current flowing into the branches. In other words, $I_{R1} + I_{R2} +$

I_{R3} = _____ = I_T. The current flowing into the branches is equal to the current flowing out from the branches. Voltmeter U_5 is measuring the voltage drop across R_3, R_4, and R_5 (the parallel section).

4. Take a good look at the current flow in the circuit. Starting from the negative terminal of the power source, I_T flows through resistor R_____ and then splits up to flow through resistors R_____, R_____, and R_____, rejoins, and then flows through resistors R_____ and R_____. Resistors R_____, R_____, and R_____ are in series with the circuit, and all of the current flows through them. Resistors R_____, R_____, and R_____ are in parallel with each other, but as a group are in series with the rest of the circuit. If one of the series resistors R_1, R_2, or R_6 were open, then there would be no current flow in the circuit because the series current path would be open.

● ***Troubleshooting Problems:***

5. Open circuit file **06-03b**. Activate the circuit and use the DMM to determine the problem. The faulty series resistor, R_____, is <u>shorted/open</u> (circle the answer) and measures _____. What about I_T? I_T measures _____, but should measure _____.

6. Open circuit file **06-03c**. Activate the circuit and determine the problem. The faulty parallel resistor, R_____, is <u>shorted/open</u> (circle the answer) and measures _____. What about I_T? I_T measures _____, but should measure _____.

7. Open circuit file **06-03d**. Activate the circuit and determine the problem. The faulty resistor, R_____, is <u>shorted/open</u> (circle the answer) and measures _____. What about I_T? I_T measures _____, but should measure _____.

8. Open circuit file **06-03e**. Activate the circuit and determine the problem. The faulty resistor, R_____, is <u>shorted/open</u> (circle the answer) and measures _____. What about I_T? I_T measures _____, but should measure _____.

9. In series-parallel circuits, you should have a rough idea of what the voltage drops are going to be and where the current is going. It is usually hard, when troubleshooting electronics equipment, to make current measurements because it is difficult to break into the circuit to insert a meter. In some equipment, provision is made to take current measurements with special current

test points. Technicians rely to a large degree on voltage drop measurements for the bulk of their troubleshooting (at least early in the course of the troubleshooting process). They place most of their emphasis on the voltage measurements and later move to somewhat more complex types of test equipment and different types of measurements.

10. Open circuit file **06-03f**. The voltage drop across the R_3-R_4 parallel circuit should measure 30 V, and it measures _____. What is wrong with this circuit? The problems are _____

_____. (This circuit has two faulty resistors.)

Activity 6.4: Loaded and Unloaded Voltage Divider Circuits

1. An unloaded **voltage divider** is a series circuit that has certain (usually according to design) voltages at the junctions between the series resistors. This type of arrangement makes up a voltage divider circuit. When loads are connected to the voltage divider, the load is in parallel (to ground) and changes the operating characteristics of the divider. This parallel load has to be taken into consideration when designing the divider to keep the voltages correct at each junction.

2. Open circuit file **06-04a**. Use the Agilent DMM and determine the voltages at TPA, TPB, and TPC? Measured voltages at TPA = _____, at TPB = _____, and at TPC = _____.

3. Open circuit file **06-04b**. This is a loaded voltage divider. Measure the voltages across the loads (R_4 and R_5) with the DMM. Measured load voltages are V_{R4} (TPC) = _____ and V_{R5} (TPB) = _____.

4. What amount of current is being drawn by the loads? Use the DMM to determine load currents. Measured load currents are I_{R4} = _____ and I_{R5} = _____.

5. Open circuit file **06-04c**. The bottom resistor (R_3) in a voltage divider is called a bleeder resistor. When designing a voltage divider, the first step is to determine bleeder current. This design starts with a bleeder current of 50 mA. Select the bleeder resistor first, which has a design requirement of 10 V at 50 mA according to the specifications. This is a simple Ohm's law calculation: R_3 = 10 V/50 mA. Calculated R_3 = _____. After completing your calculation and subsequent ones, change the resistor(s) to the calculated value(s).

6. Next, determine the value of R_2, which drops 20 V at a current of 60 mA (the 60 mA is the bleeder current or 50 mA plus the load current of 10 mA).

The voltage drop of 20 V is the voltage difference between the ends of the resistor, R_2. Calculated R_2 = _____.

7. Lastly, the value of R_1 needs to be determined. R_1 has a current of 80 mA (60 mA plus load current of 20 mA) and drops 70 V. Calculated R_1 = _____.

8. What are the wattage requirements of the three resistors R_1, R_2, and R_3? Calculated P_{R1} = _____, P_{R2} = _____, and P_{R3} = _____.

- **Troubleshooting Problems:**

9. Open circuit file **06-04d**. Yesterday, this was a perfectly running voltage divider circuit, supplying the correct voltages to the loads. Something is wrong; the voltages aren't right today. Did one of the loads change or did a divider resistor change value? Measure the voltage divider resistors and the loads. The problem is _____

_____.

10. Open circuit file **06-04e**. Now there are more problems. Earlier today this was a perfectly running voltage divider circuit, supplying the correct voltages to the loads (with a bleeder current of 100 mA). Now something is wrong; the 10 V load is smoking and the output voltages aren't right. Did one of the loads change or did something happen in the divider circuit? The problem is

_____

_____.

Activity 6.5: The Effects of Voltmeter Loading

1. One of the types of circuits that are easiest to "load down" is a voltage divider. In this exercise a voltage divider will be used to demonstrate the loading effect that a low resistance voltmeter presents to a circuit. All voltmeters have an input resistance characteristic that sometimes causes problems when they are placed in parallel with a "touchy" load such as a precision voltage divider. Usually, the more inexpensive the meter, the more potential there is for loading problems (this is typical with the inexpensive flea-market meter). And, of course, "in parallel" is the way that voltmeters are used. Most modern meters (digital types) have high input resistance ratings and do not affect a circuit to any great extent, but in the case of the "touchy" circuit or when readings are just a "little bit off," the meter always has to be taken into consideration.

2. Open circuit file **06-05a**. This circuit is supposed to provide voltages that are approximately 10 V at TPA and 30 V at TPB. Activate the circuit and notice

that the voltages are incorrect. What are the readings? The voltage at TPA is

_____ and the voltage at TPB is _____.

3. Disconnect meters U_1 and U_2 from the circuit and measure the voltages at TPA and TPB with the DMM. Now the readings are correct. Leave the DMM hooked up to TPA and reconnect U_1 and U_2. Now, the voltages become wrong again. Obviously meters U_1 and U_2 have an effect on the operation of the voltage divider, while the DMM does not seem to affect the circuit. Click on the **Settings** button on the DMM and a menu will come up to inform the viewer that the DMM resistance is 1 GΩ. Now, left-click on the panel meters U_1 and U_2. Notice that one of the meters has a low internal resistance setting.

The meter with the lowest internal resistance is U_____. It is set for

_____.

4. Any typical digital meter has an internal resistance of 10 MΩ or greater. Only inexpensive analog meters are less than that. In the case of the problem in Step 3, the internal resistance is exaggerated to show you the effect of voltmeter loading. Change the setting to 1 MΩ and check the circuit again. What are the voltage readings now? The voltage at TPA is _____V and the voltage at TPB is _____V. That is better, but if the meters are set to a higher resistance (try 10 MΩ), the voltage divider readings will be even closer.

● *Troubleshooting Problem:*

5. Open circuit file **06-05b**. The 10 V source is out of tolerance, what is wrong? The problem is _____

_____.

Activity 6.6: Dealing with Bridge Circuits

1. There is a special classification of series-parallel circuits known as **bridge circuits**. These types of circuits are often used in electronics instrumentation to interface sensors or to facilitate the measurement of natural phenomena. A typical bridge circuit is displayed in Figure 6-2.

2. Open circuit file **06-06a**. This circuit is a balanced bridge circuit (Wheatstone). The ratio of $R_1/R_2 = R_3/R_4$ has to be maintained to keep the voltmeter reading at 0 volts. R_4 is the resistor in this circuit that is usually used to measure the "unknown" resistance or phenomenon under test. The primary usage of this type of circuit is to perform tests of these types. Activate the circuit. What is the voltmeter reading? Measured voltmeter

reading = _____.

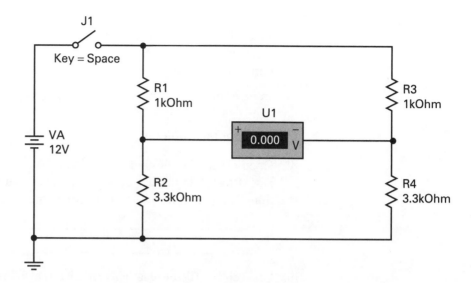

Figure 6-2 A Typical Bridge Circuit in MultiSIM

3. Change resistor values to $R_3 = 3$ kΩ and $R_4 = 9.9$ kΩ? Activate the circuit again. What happened to the voltmeter reading? The voltmeter reading

_____.

4. The ratio $R_1/R_2 = R_3/R_4$ would still be the same and the voltmeter reading should remain very close to 0 volts. Change R_4 to 10 kΩ and observe the change in the voltmeter reading. What is the reading now? The voltmeter reading with $R_4 = 10$ kΩ is _____.

5. This is the method used to balance bridge circuits: change the setting of the variable resistance to obtain the necessary balance as indicated by the meter. When the meter "zeros out," determine what the resistance of the balancing resistor (usually indicated by a dial of some sort) is to find the value of the unknown resistance.

6. Open circuit file **06-06b**. This circuit is a balanced bridge circuit with a 10-kΩ potentiometer in the place of R_4. The potentiometer is set at its 50% point and the voltmeter reading indicates an unbalanced condition. What is the "percentage" setting of the potentiometer that will restore balance to the bridge circuit so that we can determine R_____ unknown.

7. The potentiometer setting would be _____% of 10 kΩ, which is equal to _____ Ω.

8. Open circuit file **06-06c**. This bridge circuit is going to measure an unknown resistor in the range of 0 to 100 kΩ. Notice that this circuit is using an ammeter rather than a voltmeter to measure discrete differences between the two points. Both methods gives a zero indication when there is no potential difference between points A and B. Activate the circuit and toggle the keyboard key

for the potentiometer (R) until the ammeter reads 0. At that point, the resistance of the potentiometer is equal to the unknown resistor. Multiply the percentage factor to the right of the potentiometer by the value of the potentiometer itself (100 kΩ) and that is the value of the unknown resistor. The value of the unknown resistor is $R_{unknown}$ = _____.

- ● *Troubleshooting Problems:*

    9. Open circuit file **06-06d**. This circuit has lost its balance. Use the Agilent DMM to locate the defective component. You will have to disconnect portions of the circuit to be able to isolate the component fault. The problem is

    _____

    _____.

    10. Open circuit file **06-06e**. This circuit has also lost its balance. Use the Agilent DMM to locate the defective component. Again, you will have to disconnect portions of the circuit to be able to isolate the component fault. The problem

    is _____

    _____.

7. Basic Network Analysis

References

Electronics Workbench®, MultiSIM Version 7

Electronics Workbench®, MultiSIM Version 7 User's Guide

Objectives After completing this chapter the student should be able to:

- Identify a complex circuit
- Analyze complex circuits
- Simplify resistive circuits by using
  - Mesh analysis
  - Nodal analysis
  - Superposition theorem techniques
  - Thevenin's theorem
  - Norton's theorem
- Perform tee-to-pi and pi-to-tee conversions
- Troubleshoot complex circuits

Introduction

Complex circuits constitute a problem for the technician because the circuit configuration cannot be reduced to a single resistance. In the earlier circuits that have been studied (series, parallel, and series-parallel circuits), simplification techniques result in a single resistance being obtained. Complex circuits are categorized as complex because they contain more than one power source or they have other components that prevent the resistor networks from being simplified to one resistance. As stated previously, it is unusual to find simple series or parallel circuits in electronics equipment. Technicians spend the bulk of their time working on series-parallel combinations and other categories titled specifically as complex circuits. All technicians have to understand the characteristics of complex circuits to enable them to interpret schematic diagrams and to diagnose circuit faults in electronics equipment. The inability to reduce a complex circuit to a single resistor does not prevent the technician from diagnosing a circuit problem. There are many other tools, such as network theorems, that enable the circuit to be understood in its simplest terms.

Because complex circuits cannot be simplified to a single resistance for the purpose of determining total circuit load offered by the circuit, MultiSIM greatly helps engineers and electronics technicians to determine the operating characteristics of complex circuits before constructing them on a breadboard in the lab.

Activity 7.1: Recognizing a Complex Circuit

1. **Complex circuits** are easy to recognize because there are additional components or power sources that resist simplification. The circuit shown in Figure 7-1, for example, is a complex circuit. One of the most common factors indicating that a circuit is complex is that it may have two power sources.

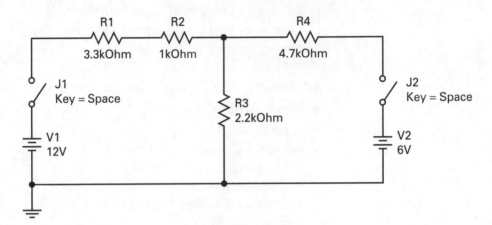

Figure 7-1 A Complex Circuit

2. Open circuit file **07-01a**. Is it possible to simplify this circuit any further than has been accomplished on this schematic? It is <u>possible/not possible</u> (circle the answer) to further simplify this circuit.

3. Open circuit file **07-01b**. This circuit is an example of how to simplify a complex circuit. Start the simplification process by combining (calculating) series resistors such as R_1 and R_2 and calling the result R_A. R_3 and R_4 would become R_B and finally, R_5 and R_6 would become R_C. What are the values of R_A, R_B, and R_C after Circuit 1 is simplified? In the simplified circuit, R_A = _____, R_B = _____, and R_C = _____.

4. Activate the circuit and determine the current value as indicated by U_1. Measured U_1 = _____.

5. Go to Circuit 2 and change the R_A, R_B, and R_C resistor values to their new calculated values. Activate both circuits and verify that the two ammeters display the same value of current for both circuits. What are the two circuit current values according to U_1 and U_2? Measured U_1 = _____ and U_2 = _____. Do U_1 and U_2 match? Yes_____ or No_____.

Activity 7.2: Mesh Analysis

1. **Mesh analysis** is a mathematical method of simplifying complex circuits. This method extends the application of Kirchhoff's voltage law.

2. Open circuit file **07-02a**. In this circuit it is necessary to write mesh equations to solve circuit parameters. Write mesh equations for the circuit and solve them for I_x and I_t. Calculated I_x = _____ and I_t = _____. A typical mesh equation for this circuit in Kirchhoff's voltage law format is shown in Figure 7-2. Kirchhoff's voltage law format shows the equality between the voltage drops and the voltage source.

Mesh 1: $3.3Ix + (2.2Ix - 2.2It) = 12$
$5.5Ix - 2.2It = 12$

Mesh 2: $1.8It + (2.2It - 2.2Ix) = -8$
$4.0It - 2.2Ix = -8$

Figure 7-2   Mesh Equations

3. These equations are the first steps of mesh analysis and they will have to be developed further to achieve the end result of determining the value of Mesh 1 (I_x) and Mesh 2 (I_t) currents. In this circuit, the mesh current I_t ends up as a negative quantity. This negative quantity indicates that the mesh current is in a direction that is opposite to the assumed direction of current for that current loop. Simply change the negative to a positive. The advantage to using MultiSIM to solve for circuit parameters in this type of circuit is that you can check out the proper answers for your calculations with a virtual circuit. If there is a numerical discrepancy, the calculations can be compared with a MultiSIM virtual circuit to isolate the problem.

4. Activate the circuit and observe U_1 and U_2 to check the validity of the mesh equations. U_3 displays the value of current flowing through R_2. What are the current flow values? Measured U_1 = _____, U_2 = _____, and U_3 = _____.

- *Troubleshooting Problems:*

5. Open circuit file **07-02b**. Use mesh analysis to determine the current flow through resistor R_1. Calculated I_{R1} = _____.

6. Now, activate the circuit and observe I_{R1} as indicated by ammeter U_1. Measured I_{R1} = _____.

7. Obviously there is a problem. What is wrong? The problem is _____
_____.

Activity 7.3: Nodal Analysis

1. **Nodal Analysis** is similar to mesh analysis, but it is based on Kirchhoff's current law Rather than Kirchhoff's voltage law to solve complex network

problems. A node is considered to be any point in a circuit where currents combine. Usually, ground is used as a reference point for solving circuit problems and making measurements. However, if ground were not present in the circuit, then an arbitrary reference point would be used for voltage measurements.

2. Open circuit file **07-03**. Write a nodal equation for the circuit and solve it for V_x, which represents the unknown voltage drop across R_3. Calculated V_x = _____. Remember that $V_x = V_{R3}$.

3. Activate the circuit and use the Agilent DMM to measure the voltage drop across resistor R_3. Measured V_{R3} = _____.

Activity 7.4: The Superposition Theorem

1. The **superposition theorem** is another valuable technique for analyzing complex circuits that have more than one power source. With this method, you act as if one of the power sources is a short determining current values and voltage drops in reference to that one power source. Then you proceed to the other power source and act as if the first power source is a short, again determining current values and voltage drops in reference to the second power source. Now, algebraically combine the results to determine circuit parameters.

2. Open circuit file **07-04a**. There are three circuits on the workspace. Circuit 1 represents the circuit that power source V_1 sees, Circuit 2 represents the circuit that power source V_2 sees, and Circuit 3 represents both circuits combined with ammeters in each current path.

3. Use the superposition theorem to calculate I_{R3}. Calculated I_{R3} = _____.

4. Activate the circuit and verify the calculation with the U_3 display. Measured I_{R3} = _____.

5. Open circuit file **07-04b**. Use the superposition theorem to calculate circuit operation. First, consider V_2 to be a short and calculate the circuit parameters. Then consider V_1 to be a short and recalculate the circuit parameters again. Finally, algebraically superimpose the results on each other. Use the multimeter to verify superposition results by measuring voltage drops and branch currents, particularly I_{R3}. Calculated I_{R3} = _____.

6. Now, activate Circuit 1 to verify the calculations. Measured I_{R3} = _____.

● *Troubleshooting Problem:*

7. Open circuit file **07-04c**. This circuit has been returned from the field and you, as a factory service technician, have to find the problem and report

your findings. The circuit is the same circuit you tested in Steps 5 and 6. Activate the circuit. Are the results the same as for Step 6 above? Obviously there is a problem. Determine what the problem is. The problem is _____

_____.

Activity 7.5: Using Thevenin's Theorem

1. With **Thevenin's theorem**, simple resistive circuits or portions of a more complex circuit can be converted into a simple equivalent circuit. A typical Thevenin's theorem calculation results in a simplified series circuit with a constant voltage source identified as V_{TH} and a single resistance identified as R_{TH}. In a circuit where only a portion of the circuit is being simplified using Thevenin's theorem, the point where the simplification is taking place is called the point of simplification. See Figure 7-3 for an example of a circuit simplified according to Thevenin's theorem.

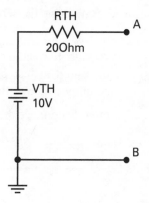

Figure 7-3   Thevenin's Theorem Equivalent Circuit

2. As can be seen in Figure 7-3, a thevenized circuit replaces a resistive circuit with a two-terminal equivalent circuit. This simplified circuit consists of a single voltage source called V_{TH} and a series resistance called R_{TH}. V_{TH} is the open terminal voltage across terminals A and B. R_{TH} is the single resistance in series with the voltage source and is also in series with any load attached to terminals A and B.

3. Open circuit file **07-05a**. Go through the steps involved in determining V_{TH} and R_{TH}. Circuit A is the beginning circuit. Circuit B represents the first stage of thevenizing the circuit, obtaining the resistance of parallel components R_2 and R_3. Calculated R_2–R_3 = _____.

4. Circuit C displays the value of R_{TH} and the power source that is yet to be determined by further analysis. Circuit D displays Circuit B with $V_1 = 12$ V. Circuit E ties it all together with the calculated results of V_{TH} and R_{TH} being displayed. Calculated V_{TH} = _____V and R_{TH} = _____.

5. Use the DMM and measure the circuit to compare the measured values of V_{TH} and R_{TH} with the calculated values. To obtain V_{TH} and R_{TH}, use Circuit B for the resistance measurement and Circuit D for the voltage measurement.

    Measured V_{TH} = _____V and R_{TH} = _____.

6. One important use of Thevenin's theorem is in determining how much current will flow through a load connected to output terminals A and B in the thevenized circuit.

7. Open circuit file **07-05b**. In this circuit, R_1 is to be attached to the circuit as the load. As can be seen, this is a simple series circuit and it should be easy to determine the voltage drop across the load (R_1) as well as the current flowing through it. Use the DMM to determine those quantities. Measured I_{LOAD} = _____ and V_{LOAD} = _____.

- ● *Troubleshooting Problem:*

8. Open circuit file **07-05c**. Calculate V_{TH} and R_{TH}. Calculated V_{TH} = _____V and R_{TH} = _____. Now, measure V_{TH} and R_{TH}. Remember, V_{TH} is the voltage that is measured at test points A and B. R_{TH} is the resistance measured at test points A and B with the power supply shorted (connect a jumper across the power supply from test point C to test point D and then disconnect the power source). Measured V_{TH} = _____ and R_{TH} = _____. Obviously, there is a problem. What is it? The problem is

    _____.

Activity 7.6: Using Norton's Theorem

1. **Norton's theorem** is similar to Thevenin's theorem except that the final result is an equivalent circuit consisting of a current source in parallel with a resistance rather than a voltage source in series with a resistance. Norton's theorem uses a current-related approach while Thevenin's theorem uses a voltage-related approach. A constant current source provides a current output that is unaffected by external circuit resistance. No matter what the load does, the output current remains the same. Figure 7-4 displays the MultiSIM symbol for a constant current source, and Figure 7-5 displays a Norton's theorem equivalent circuit.

Figure 7-4   MultiSIM Symbol for a Constant Current Source

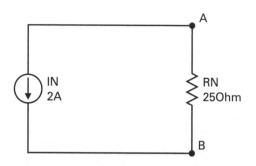

Figure 7-5   Norton's Theorem Equivalent Circuit

2. Norton's theorem uses the same approach to R_{TH} as Thevenin's theorem and, thus, R_{TH} is the same as R_N. This resistance has the same value for both theorems, $R_N = R_{TH}$. The application of the results is different with Norton's theorem; R_N is in parallel with the current source I_N and with an applied load. The Norton current divides between the load and R_N.

3. Open circuit file **07-06**. Circuit 1 represents the initial circuit. Start by determining the value of R_N. Calculated R_N = _____. Circuit 2 represents the thevenized circuit that displays a value for R_N.

4. To calculate I_N, determine the amount of current that would be flowing through a shorted output from A to B which, essentially, shorts out R_2. This leaves R_1 in series with the voltage source, a simple series circuit. Calculated I_N = _____ mA. Change the values in Circuit 4 to represent the calculated values of R_N and I_N.

5. Connect R_{LOAD} to terminals A and B of Circuit 4. Predict the current through the load and the voltage across it. The Norton's current will divide proportionally between the load resistor and R_N. The voltage across the load resistor (and R_N) will be equal to V_{TH}. Predicted I_{LOAD} = _____.

6. Now measure I_{LOAD}. Measured I_{LOAD} = _____.

Activity 7.7: Tee-to-Pi and Pi-to-Tee Conversions

1. **Tee circuits** and **pi circuits** are also referred to in the electrical industry as **wye** and **delta circuits**. The terms tee and pi will be used in this study. See Figures 7-6 and 7-7 on page 70 for examples of tee and pi circuits. The purpose of the conversion process is to attain a circuit in one of the two configurations (tee or pi) that is electrically equivalent to the other configuration. The term wye is equivalent to tee, and the term delta is equivalent to pi throughout the electrical/electronics industry.

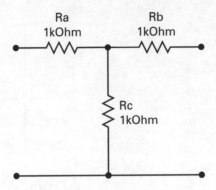

Figure 7-6   Tee Configuration

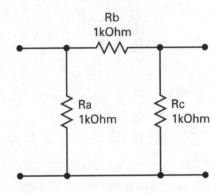

Figure 7-7   Pi Configuration

2. The first conversion to be accomplished is the tee-to-pi conversion. Open circuit file **07-07a**. In this circuit file, Circuit 1 is a tee circuit and Circuit 2 is a pi circuit. Activate the circuits and notice that the pi circuit draws almost three times more current from the source than the tee circuit and provides only about one-third more output voltage and current. Enter the data from both circuits in Table 7-1.

3. Using the tee-to-pi formulas, change the pi circuit so that it is electrically equivalent to the tee circuit. The goal in this conversion is to find out what values in the pi circuit would provide the same load to the voltage source as the tee circuit with the same output to the load. Calculated $R_{ab} =$ _____,

    $R_{ac} =$ _____, and $R_{bc} =$ _____.

4. Change the values of R_{ab}, R_{ac}, and R_{bc} in Circuit 2 to match the calculations. Activate the circuit and enter the revised Circuit 2 data in Table 7-1 (the Circuit 1 data will be the same as previously entered). Notice that, by changing the resistors in Circuit 2, the inputs and outputs are now identical for both circuits. That is the purpose of the conversion.

5. The second conversion under consideration is the pi-to-tee conversion. Open circuit file **07-07b**. In this circuit file, Circuit 1 is a pi circuit and

	$R_a = R_b = R_c = 1 \text{ k}\Omega$ (The Original Configuration)		$R_{ab} = R_{ac} = R_{bc} = ? \text{ k}\Omega$ (The Modified Configuration)	
	Source 1	Source 2	Source 1	Source2
Source Voltage				
Source Current				
	Output 1	Output 2	Output 1	Output 2
Output Voltage				
Output Current				

Table 7-1 Tee-to-Pi Conversion

Circuit 2 is a tee circuit. Activate the circuits and notice that the tee circuit draws about one-third the current of the pi circuit from the source and provides less output voltage and current. Enter your data from both of the circuits in Table 7-2.

	$R_{ab} = R_{ac} = R_{bc} = 1 \text{ k}\Omega$ (The Original Configuration)		$R_a = R_b = R_c = ? \text{ k}\Omega$ (The Modified Configuration)	
	Source 1	Source 2	Source 1	Source2
Source Voltage				
Source Current				
	Output 1	Output 2	Output 1	Output 2
Output Voltage				
Output Current				

Table 7-2 Pi-to-Tee Conversion

6. Using the pi-to-tee formulas, change the tee circuit so that it is electrically equivalent to the pi circuit. The goal in this conversion is to find out what values of resistance the resistors in the tee circuit need to be in order to provide the same load to the voltage source as the pi circuit and with the same

output to the load. The new values are $R_a =$ _____, $R_b =$ _____, and $R_c =$ _____.

7. Change the values of R_a, R_b, and R_c in Circuit 2 to match the calculations. Activate the circuits and enter the revised Circuit 2 data in Table 7-2 (the Circuit 1 data would be the same in both cases). Notice that by changing the resistors in Circuit 2, the inputs and outputs are identical for both circuits. This is the purpose of the conversion.

8. Open circuit file **07-07c**. Convert this tee circuit to an equivalent pi circuit and enter your calculations in Table 7-3. Calculated $R_{ab} =$ _____, $R_{ac} =$ _____, and $R_{bc} =$ _____.

9. Change the value of the Circuit 2 components to match the converted data from Circuit 1. Activate the circuit, determine equivalency, and enter your data in Table 7-3.

	The Original Configuration		**The Modified Configuration**	
	Source 1	**Source 2**	**Source 1**	**Source2**
Source Voltage				
Source Current				
	Output 1	**Output 2**	**Output 1**	**Output 2**
Output Voltage				
Output Current				

Table 7-3   Tee-to-Pi Conversion Practice Problem

10. Open circuit file **07-07d**. Convert this pi circuit to an equivalent tee circuit and enter the data in Table 7-4. Calculated $R_{ab} =$ _____, $R_{ac} =$ _____, and $R_{bc} =$ _____.

11. Again, note that the original configuration data for Circuit 2 has no relationship to Circuit 1 at this point. Change the resistor values of Circuit 2 to make it electrically equivalent to Circuit 1.

	The Original Configuration		The Modified Configuration	
	Source 1	Source 2	Source 1	Source2
Source Voltage				
Source Current				
	Output 1	Output 2	Output 1	Output 2
Output Voltage				
Output Current				

Table 7-4   Pi-to-Tee Conversion Practice Problem

8. Electrical Power Sources

References

Electronics Workbench®, MultiSIM Version 7

Electronics Workbench®, MultiSIM Version 7 User's Guide

Objectives After completing this chapter the student should be able to:

- Connect and use power sources found in MultiSIM
- Determine internal resistance of DC power sources
- Connect DC power sources in series
- Connect DC power sources in parallel
- Analyze additional DC power sources found in MultiSIM
- Discuss and analyze AC power sources

Introduction

There are many methods used by industrialized nations to obtain electrical energy for industry, home, and commerce. In electronics equipment, all methods are used to one degree or another; but basically in the generation of power for electrical and electronic circuits, the primary methods of power generation are by generators (electromagnetic means) or by the use of cells and batteries (chemical means).

Figure 8-1 displays symbols representing some of the voltage sources available in MultiSIM. Some of these symbols have already been discussed in Chapter 1, but they will be briefly covered again. There are other power sources available in the MultiSIM menu, selections that are less commonly used than those shown in Figure 8-1. The other means of power generation such as static electricity, mechanical means (piezoelectric), solar means (photovoltaic), and heat generation (thermoelectric) that are viable power sources in industry are not found in MultiSIM.

Among the various DC power sources used in MultiSIM, the source represented by a battery symbol is the source that has been used in the associated MultiSIM circuit files for this text up to this point. The battery symbol is always used when there is a need for DC voltage with a few exceptions such as V_{CC}, V_{DD}, V_{SS}, and V_{EE}. The V_{CC}, V_{DD}, V_{SS}, and V_{EE} source symbols have fixed meanings in industry, but in MultiSIM they represent user-defined voltages. For instance, V_{CC} could be defined as +5 VDC for a particular circuit and −5 VDC for another circuit. In either case, wherever the V_{CC} symbol is placed within that particular circuit, it represents only the defined voltage.

In Figure 8-2, the box with the +12 VDC caption is a subcircuit symbol. The subcircuit symbol is frequently used in MultiSIM to place a larger circuit inside a

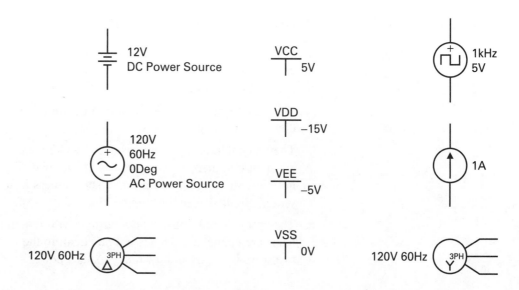

Figure 8-1   Some Voltage Sources Used in MultiSIM

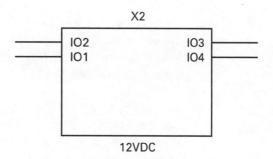

12VDC

Figure 8-2   The MultiSIM Subcircuit Symbol

box, with only outputs available. This is a valuable tool for decreasing overall circuit size on the workspace and for presenting function block troubleshooting. You will study the subcircuit in Activity 8.5.

Some other MultiSIM DC power sources include the variable square wave signal generator, a type of pulsating voltage source, and the DC current source displayed as a circle with an arrow pointing upward.

Other power sources in MultiSIM include the **alternating current** (AC) power source symbol, which represents an alternating current signal, as well as additional power sources with variable voltage and variable frequency capabilities. All of these AC signals can also be obtained from the function generators, which are pieces of test equipment that are kept in the **Instruments** library. Activate MultiSIM and click on the **Instruments** menu bar; notice that the second instrument from the left is the function generator. Drag it out and take a look at it. This instrument will be used extensively in the chapters to come. Also, there is an Agilent function generator, the blue instrument fourth from the right end of the bar.

Activity 8.1: DC Power Sources and Internal Resistance

1. The power sources used in MultiSIM are ideal power sources; they have no internal resistance. Ideal (virtual) power sources have some characteristics that "real" power sources do not have. For example:

   a. They do not wear out over a period of time like a battery or even an AC power source does.

   b. They do not have internal resistance, which means that they can provide a constant output voltage regardless of the load. This means that they will not blow fuses or trip circuit breakers (unless a fuse is designed into the virtual circuit) when they are overloaded.

   c. They provide unlimited output current with few limitations on how many loads are connected. This is not practical; in the real world it is necessary to always be aware of how many loads a power source can handle.

2. To represent a "real" DC power source (a battery) with internal resistance, a low value resistor is placed in series with the output terminal to simulate the internal resistance of a source.

3. Open circuit file **08-01a**. Activate the circuit and observe that the output of the power source as displayed on U_3 is equal to the power source voltage of 100 V. In this circuit, R_{int} represents the internal resistance of the power source, a resistance of 1 Ω.

4. Activate S_1 to apply a load (R_2) to the circuit. At this point, the load is 100 Ω and the internal resistance of the power source is only 1% of the load. This should result in about a 1% voltage drop across the internal resistance. U_1 represents the drop across the simulated internal resistance, U_2 represents the actual voltage of the power source, and U_3 represents the percentage of the voltage being applied to the load. How much voltage is being dropped by the simulated internal resistance of the power supply (as indicated by U_1)?

   Measured R_{int} = _____.

5. Toggle the potentiometer (R_2) in 1% intervals and notice the drop in the voltage being applied to the load as the internal resistance drops more of the voltage. Fill in Table 8-1 with the test data.

6. The voltage across the load will equal one-half of the rated output of the power source (100 V in this case) when the potentiometer is equal to the internal resistance. What is the percentage setting of the potentiometer when

   U_3 = 50 V? The potentiometer is at _____% when U_3 is at 50 V. This indicates that the internal resistance of the power supply can be calculated.

   Calculated R_{int} = _____.

Load	U1 Reading	U2 Reading	U3 Reading
% and Ω	$R_{internal}$ V	Source V	(Output Voltage)
No Load	0 V	100 V	100 V
100%-100 Ω			
90%-90 Ω			
75%-75 Ω			
50%-50 Ω			
25%-25 Ω			
10%-10 Ω			
5%-5 Ω			
4%-4 Ω			
3%-3 Ω			
2%-2 Ω			
1%-1 Ω			
0%-0 Ω			

Table 8-1   Power Source Loading Versus Output Voltage

● *Troubleshooting Problems:*

7. Open circuit file **08-01b**. This power source (Subcircuit X1) has a problem; it has too much internal resistance. Calculate the value of the internal resistance. Calculated R_{int} = _____.

8. Open circuit file **08-01c**. This circuit is not operating properly. Use the Agilent DMM to troubleshoot the circuit. The problem is: _____

_____.

Activity 8.2: DC Power Sources in Series

1. The voltage outputs of DC power sources that are placed in a series configuration always provide a voltage that is the algebraic sum of the power sources. This includes power sources that aid as well as oppose one another. In MultiSIM, it is easy to change the voltage output of a power source. If more than one specific voltage is needed, then it is possible to stack power sources

in series with voltage outputs between the individual power sources or use independent sources with different voltage outputs.

2. Open circuit file **08-02a**. This file has several circuits with various combinations of stacked power sources. What is the calculated voltage output for each of them? Use the DMM to measure the voltage outputs and then record the data in Table 8-2.

Circuit Number	Calculated DC Output	Measured DC Output
Circuit 1		
Circuit 2		
Circuit 3		
Circuit 4		

Table 8-2 DC Power Sources: Series-Aiding and Series-Opposing

● *Troubleshooting Problems:*

3. Open circuit files **08-02b**, **08-02c**, **08-02d**, and **08-02e**. There is a problem with each of these power supply circuits. Calculate the correct voltage outputs for each circuit and then measure the actual outputs. Try to determine what is wrong with each power supply circuit. The calculated voltage output for each circuit is: Circuit 08-02b _____, Circuit 08-02c _____, Circuit 08-02d _____, and Circuit 08-02e _____.

4. The individual problem with each of the circuits is:

Circuit 08-02b _____.

Circuit 08-02c _____.

Circuit 08-02d _____.

Circuit 08-02e _____.

Activity 8.3: DC Power Sources in Parallel

1. When "real" DC power sources are placed in parallel (with equal voltage outputs), the output voltage output is fixed. The current output is theoretically equal to the sum of the individual currents of the parallel power sources. There are problems with this type of hookup and it is not generally recommended, but it can be done if necessary. The reason generally given for such a maneuver would be to provide more current than an individual source is able to provide. This type of solution, the parallel connection of power sources, is often seen in automotive applications where more than one battery

is needed to provide power for auxiliary equipment. This hookup is more complicated than simply connecting the batteries in parallel.

2. There should be no need to use parallel power sources in MultiSIM because the virtual power sources have no current limitations. In fact, MultiSIM will give an alarm condition if an attempt is made to parallel power sources.

3. Open circuit file **08-03a**. Activate the circuit and a MultiSIM alarm condition will be triggered by this method of power supply connection. MultiSIM power sources can be connected in parallel only if a resistor is placed in series with each of the outputs. Place the 1 Ω resistors in series with the power sources and reactivate the circuit. All should be well now. Did the circuit operate properly? Yes_____ or No_____.

4. With the resistors still in the circuit, change the voltage setting of power source V_1 to 6 V. Activate the circuit, and the output voltage displayed on the meter changes to 9 V. This is halfway between 12 V and 6 V; each of the resistors has a voltage drop. Use the DMM to measure the voltage drops across the 1 Ω resistor. Measured V_{R2} = _____ and V_{R3} = _____.

5. Open circuit file **08-03b**. Calculate the correct output voltage according to the schematic. The calculated circuit output voltage as indicated by U_1 should be: Calculated Output Voltage = _____.

6. Activate the circuit and observe circuit operation. The measured voltage is: Measured U_1 = _____.

7. Obviously, there is something wrong with the circuit. Use the DMM to find the problem. The problem is: _____
_____.

Activity 8.4: More DC Power Sources in MultiSIM

1. This activity introduces some additional MultiSIM DC power sources. The V_{CC}, V_{DD}, V_{EE}, and V_{SS} sources are under consideration. These additional power sources are displayed in Figure 8-3. These types of power sources would primarily be used in digital and solid state circuits.

VCC	VDD	VEE	VSS
5V	5V	–5V	0V

Figure 8-3   Additional DC Power Sources

2. Open circuit file **08-04**. Measure each of the voltage sources to determine their default values. Measured V_{CC} = _____, V_{DD} = _____, V_{EE} = _____, and V_{SS} = _____.

3. These default voltages can be changed. Go to the menu for each power source and change the default voltage settings to: $V_{CC} = +5$ VDC, $V_{DD} = -5$VDC, $V_{EE} = 12$ VDC, and $V_{SS} = -12$ VDC. Use the DMM to verify the new settings. Measured $V_{CC} =$ _____, $V_{DD} =$ _____, $V_{EE} =$ _____, and $V_{SS} =$ _____.

4. There is one more source in MultiSIM that, technically, can be designated as a DC source. It is the square wave generator. This generator puts out a signal that can be adjusted to rise and fall from 0 V and never go below the 0 V point. Although this voltage acts like AC, it is truly DC. This generator will be used in a later chapter.

Activity 8.5: The Subcircuit as a Power Source

1. Open circuit file **08-05**. This circuit has a 12 VDC subcircuit connected to external components (J1 and X1). Activate the circuit, and you will observe the lamp turning on when you actuate Switch J1. You can see the contents of the subcircuit by double left-clicking on the rectangle representing the subcircuit. The circuit contents consist of _____ _____.

Activity 8.6: Introducing the Alternating Current (AC) Power Source

1. Open circuit file **08-06**. An Agilent DMM is set up to measure the AC output of the AC power source as seen in Figure 8-3. Activate the circuit and determine the output of the power source at point A. Notice that the DMM is set for AC rather than DC. The output voltage of the AC power source output = _____.

2. In AC circuits, voltage is represented by VAC, which might best be defined as **"Volts under Alternating Current conditions."**

9. Direct Current Test Equipment

References

Electronics Workbench®, MultiSIM Version 7

Electronics Workbench®, MultiSIM Version 7 User's Guide

Objectives After completing this chapter the student should be able to:

- Connect voltmeters, ammeters, and ohmmeters in DC and AC circuits
- Determine and adjust the internal resistance of current and voltage meters found in MultiSIM
- Understand the purpose of controls/connectors/software settings for MultiSIM test instruments
- Calculate the value of shunt resistors for ammeters
- Calculate the value of multiplier resistors for voltmeters
- Use the oscilloscope as a DC measuring instrument

Introduction

Voltmeters, ohmmeters, ammeters, and oscilloscopes are the eyes and ears of the electronics technician whenever he or she attempts to work on the many types of electronics equipment needing service today. Whatever can be measured in electronics becomes part of the technical knowledge base that is so helpful and necessary concerning the physical operation of electronics equipment under repair or in need of adjustment.

In this study, because of the lack of hands-on capabilities, it is not possible to study electrical/electronic instruments or meters of the analog type, which have a moving-coil movement. These analog types of meters have to be physically present to be able to appreciate their characteristics. Also, they are not as prevalent in their usage as they were in the past. The types of meters that will be studied are digital; they are virtual instruments, easy to use and simulate in a software program.

The technician not only has to know how to use the electronics test equipment available, but also whether a specific reading obtained with a particular piece of test equipment can be trusted. If the technician jumps to a wrong conclusion as the result of erroneous readings and/or faulty test equipment, he or she can waste a lot of time instead of promptly completing the task at hand. The technician should have a certain amount of general knowledge about the test equipment he or she is using in order to recognize false alarms and to prevent the subsequent loss of valuable time.

Activity 9.1: Using Digital Multimeters (DMMs) in DC Circuits

1. The voltmeters present in MultiSIM are of three types: the generic Digital Multimeter (DMM) and the Agilent Multimeter from the **Instruments** menu, and the panel type of digital meter found in the **Indicators** menu. In this activity there will be further study of the DC applications of the DMMs, which have already been used extensively in previous activities. The faceplates of the generic MultiSIM and the Agilent multimeters are shown in Figure 9-1. The various measurement options will be discussed in the following activities.

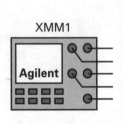

XMM1

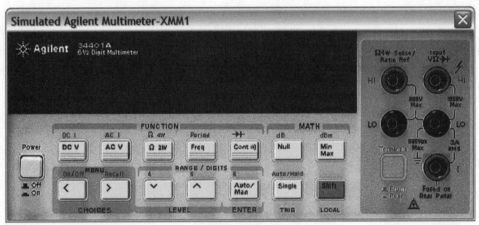

XMM2

Figure 9-1   MultiSIM Digital Multimeters

2. When using the ammeter (current flow) function of the generic DMM, it is necessary to left-click the "**A**" function and the proper signal mode (DC or AC). When connecting the DMM into the circuit to measure current flow, use the two connection points, at the bottom of the meter, marked "+" and "–". These connection points have polarity and have to be connected correctly

when measuring quantities requiring positive and negative orientation (DC voltages and currents). For nonpolarized quantities, such as resistance and AC quantities, the negative and positive orientation does not matter.

3. When using the ammeter (current flow) function of the Agilent DMM, it is necessary to left-click on **Power**, then left-click on the **Shift** button at the lower right of the meter face, and finally to left-click on **DC V** or **AC V** depending on whether you are reading DC or AC amperage values. When connecting the DMM into the circuit to measure current flow, use the two connection points at the lower right of the meter marked "3A" (red) and "LO" (black). These connection points have polarity and have to be connected correctly when measuring quantities requiring positive and negative orientation (DC currents). For nonpolarized quantities, such as AC current, the negative and positive orientation does not matter.

4. Open circuit file **09-01a**. In this circuit both of the DMMs are connected (in series) in the circuit to measure DC current shown in Figure 9-2. Notice the progression of current flow from the negative terminal of the power source, into one end of the resistor; out of the other end of the resistor into the negative terminal of the first meter, out the positive terminal of the meter, through

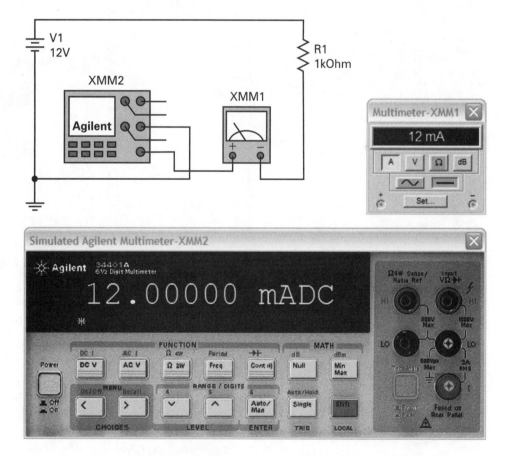

Figure 9-2 Digital Multimeter Connection to Measure DC Current

the next meter in a similar fashion, and then to the positive terminal of the power source, which completes the series circuit path. Activate the circuit and determine the current reading. Measured DC current = _____.

5. The DC voltmeter function of the digital multimeter will be discussed next. Open circuit file **09-01b**. This circuit is connected to measure DC voltage. Notice the meter is now connected in parallel with the component under test, resistor R_1. Current flows through the component, and the voltmeter displays the amount of voltage the resistor is dropping in the circuit (see Figure 9-3). Activate the circuit and determine the voltage reading. The measured DC voltage reading is _____.

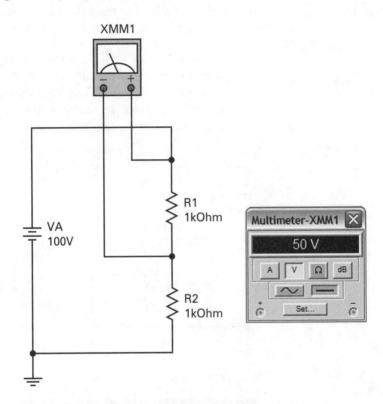

Figure 9-3   Another Digital Multimeter Connection to Measure DC Voltage

6. Now, take a look at the ohmmeter function of the digital multimeter.

   a. Open circuit file **09-01c1** (see Figure 9-4). In this circuit, the meter is connected to measure resistance. Notice the meter is measuring a resistor that is part of a circuit with the DC power source disconnected. It is necessary to disconnect the power source whenever a resistance measurement is being made. This has to be done in MultiSIM, which is simulating real-world activity, as well as in the real world where protection of the meter is of prime importance. Activate the circuit and determine the resistance of R_1. The measured resistance of R_1 is _____.

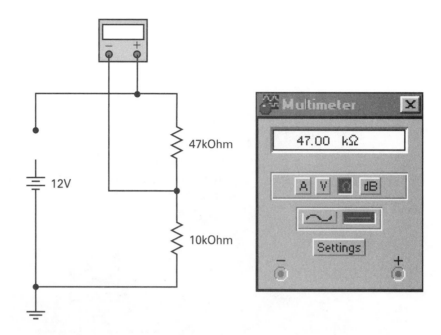

Figure 9-4   DMM Connection to Measure In-Circuit Resistance

b. Open circuit file **09-01c2** (see Figure 9-5). The meter is connected to measure resistance. In this case, notice that the meter is measuring a resistor "out of circuit." Activate the circuit. What is the resistance reading?

The measured resistance is _____.

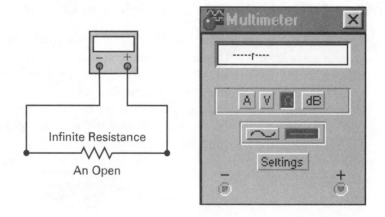

Figure 9-5   DMM Connection to Measure Infinite Resistance:
"An Open"

c. Notice the "**i**" on the display of the DMM; this indicates infinite resistance. Electronics technicians usually refer to a measurement with infinite resistance as being an "**open**" circuit. For example, if a wire is broken between two points in a circuit, that point of breakage is referred to as an open circuit.

d. Open circuit files **09-01c3** (see Figure 9-6). The meter is connected to measure resistance. Again, notice the meter is measuring a resistor "out of circuit." Activate the circuit. What is the resistance reading? The measured

resistance is _____.

e. Electronic technicians usually refer to zero ohms as being a "**short**" when the measurement is a fault condition. A short can be further defined as any place in a circuit where two or more points are making an unwanted connection.

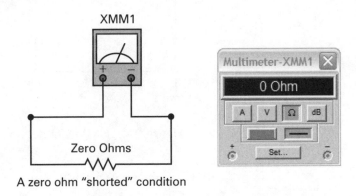

Figure 9-6   DMM Connection to Measure Zero Ohms: "A Short"

Activity 9.2: Using Digital Panel Meters in DC Circuits

1. The MultiSIM digital panel meters are located in the **Indicators** menu on the taskbar. There are two basic types: the ammeter and the voltmeter, each of which has four views depending on the positive/negative and vertical/ horizontal orientation necessary for a specific circuit application. These meters can be used for DC or AC measurements. Now the DC function will be discussed. Figure 9-7 presents four of the meters used in the MultiSIM program.

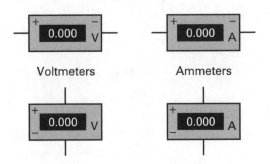

Figure 9-7   Some Digital Panel Meters Found in MultiSIM

2. The digital ammeter works essentially the same as the ammeter function on the generic DMM that you have previously used. One advantage to this type of meter is that as many of these panel meters as desired can be brought out onto the workspace. This also can be done with the DMMs. One disadvantage of this type of digital panel meter is that its only function is to measure current; it does not measure voltage or resistance.

3. Open circuit file **09-02a**. This circuit is ready for the installation of an ammeter between TPA and TPB. Go to the **Indicators** menu, open it, and drag an ammeter onto the workspace. Make sure that you select a meter that has the correct orientation and polarity to install in the circuit (Ammeter_V). Install the ammeter in the circuit between TPA and TPB. Activate the circuit and note

   the meter reading. The measured DC current reading is _____.

4. The digital voltmeter works essentially the same as the voltmeter function on the generic DMM that you have previously used. Again, as with the digital ammeter, this meter is a single-function meter.

5. Open circuit file **09-02b**. This circuit is ready for a digital voltmeter to be connected across R_2 by attaching the meter leads to TPA and TPB. Go to the **Indicators** menu, open it, and drag a voltmeter onto the workspace. Make sure that you select a meter that has the correct orientation and polarity to install in the circuit (Voltmeter_V). Now connect the voltmeter to TPA and TPB in the circuit. Activate the circuit and note the meter reading. The

   measured DC voltage reading across R_2 is _____.

Activity 9.3: Internal Resistance of Current and Voltage Meters

1. Meters can offer a load problem for electronic circuits, as was previously covered in Chapter 6, Activity 6.5. In a high-resistance circuit, a low-resistance voltmeter can adversely affect a circuit when placed in parallel with circuit components; it provides an additional parallel path for current flow. A high-resistance ammeter can also adversely affect a circuit under test by placing additional and undesired series resistance into a low-resistance circuit. Either way, internal meter resistance always has to be taken into account whenever electronics circuits are being tested.

2. The digital panel meter internal resistance can be changed by double left-clicking on the meter, clicking on the **Value** tab, and setting the **Resistance (R)** to the desired value. In a high resistance circuit 1 GΩ is the best setting.

3. Open circuit file **09-03a**. In this high-resistance circuit, two types of voltmeters are being used to measure the voltage drop across the same resistor.

   Calculate the voltage drop across R_2. Calculated V_{R2} = _____.

4. Activate the circuit and note the voltage readings. What are they? Measured

   V_{R2} = _____.

5. There is an obvious discrepancy. Check the internal resistance setting of the digital panel meter. What is wrong, and what can be done about it? The problem is _____ and it can be corrected by _____. Correct the internal resistance setting of U_1 by setting it to 1 GΩ. Activate the circuit and verify that the voltage reading is now the same as the calculation in Step 3.

6. Move the mouse pointer to the generic DMM and left-click on the **Settings** block. Notice that the voltmeter resistance setting is set to 1 GΩ. Change the setting to 1 MΩ and reactivate the circuit. Now the readings are wrong again. Obviously, the internal resistance setting is important in high-resistance circuits. Reset the DMM internal resistance back to 1 GΩ. What this means is that when the meter with an internal resistance of 1 MΩ is placed in parallel with the circuit resistor of 1 MΩ, a parallel path for current flow is created. The two resistances are of equal value and half of the current will flow through each parallel path, thereby changing the voltage drop across the resistor.

● **Troubleshooting Problems:**

7. Open circuit file **09-03b**. In this circuit, two types of MultiSIM ammeters are being used to measure the current flow through a 0.5-Ω resistor. Calculate current flow through R_1. Calculated V_{R1} = _____.

8. Activate the circuits and note the current readings. What are they? The DMM current reading is _____. The U_1 (digital panel meter) current reading is _____.

9. There is an obvious discrepancy. What is wrong? What can be done about it? The problem is _____

_____

and it can be corrected by _____

_____.

Correct the problem by replacing U_1. Reactivate the circuit and verify that current readings are correct.

Activity 9.4: Calculating the Value of Shunt Resistors for Ammeters

1. The movement of an analog ammeter is rated in "the amount of current that will produce full-scale deflection of the needle," a visible left-to-right movement that indicates, in a linear fashion (generally), how much current is flowing through the meter. This current flow cannot be more than the amount that is necessary to produce full-scale deflection, or (potentially) damage to the meter movement could result. The excess current, when it is necessary to

measure more than rated, full-scale current, has to be bypassed around the meter through a "**shunt**" resistor. The basic circuit for an ammeter shunt is shown in Figure 9-8. For example, if a specific ammeter has an internal resistance of 100 Ω for a full-scale needle deflection current of 1 mA, and 100 mA is being measured, then it is necessary to make sure that 99 mA bypasses the meter coil to prevent damage.

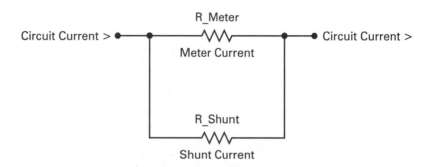

Figure 9-8   An Ammeter Shunt Resistor Circuit

2. Open circuit file **09-04**. This circuit has a meter movement rated at 1 mA for full-scale deflection of the meter and measures 50 Ω of resistance. It is desired to measure 100 mA and have full-scale deflection. What value of shunt resistor would bypass the extra current? Use the formula, $R_{SHUNT} = I_{METER} \times R_{METER}/I_{SHUNT}$. The calculated value of the shunt resistor is

_____.

3. Change the value of the shunt resistor, R_{SHUNT}, to the calculated value. Activate the circuit and verify that the calculations are correct. The meter

current is _____. The shunt current is _____.

4. U_1, representing the current through the meter movement should have 1 mA flowing through it, and U_2, representing the current through the meter shunt resistor, should have 99 mA of current flowing through it.

Activity 9.5: Calculating the Value of Multiplier Resistors for Voltmeters

1. The movement of an analog voltmeter is rated in the same manner as an ammeter because the basic meter movement reacts to current flow through it. The difference between the ammeter circuit and the voltmeter circuit is that, instead of bypassing potentially damaging current around the meter as with ammeters, "**multiplier**" resistors are placed in series with the voltmeter meter movement to drop excess voltage and to keep the current in the proper range. The basic circuit for a voltmeter is shown in Figure 9-9 on page 90. For example, if a specific voltmeter has an internal resistance of 100 Ω for a full-scale needle deflection current of 1 mA, the meter would drop 100 mV. When measuring 10 V and it is necessary to obtain full-scale deflection at that voltage, then the multiplier resistor has to drop 9.9 V.

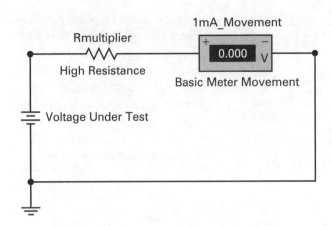

Figure 9-9   A Voltmeter Multiplier Resistor Circuit

2. Open circuit file **09-05**. This circuit has a meter movement rated at 1 mA for full-scale deflection and measures 50 Ω of resistance. The voltage under test is 100 V and that should cause full-scale deflection. What value of multiplier resistor would drop the extra voltage? Use the formula $R_{MULTIPLIER} = (V_{RANGE}/I_{FULL-SCALE}) - R_{METER}$. The calculated value of the multiplier resistor is _____.

3. Change the value of the multiplier resistor, $R_{MULTIPLIER}$, to the calculated value. Activate the circuit and verify that the calculations are correct. U_1 displays the voltage drop across the multiplier resistor, U_2 displays the voltage drop across the meter movement (it should be 50 mV) and U_3 displays the current flowing through the meter (it should be 1 mA). The excess voltage should be dropped across the multiplier resistor. V_{U1} = _____. V_{U2} = _____. Meter current (I_{U3}) = _____.

Activity 9.6: The Oscilloscope as a DC Test Instrument

1. The **Oscilloscope** (scope or o-scope) is one of the most widely used and capable test instruments available to the electronics technician. It can be used to measure DC as well as AC, but is most commonly used for its AC measurement capabilities. However, it will measure DC very well. When you first look at an oscilloscope you might be overwhelmed with all of the switches and controls (see Figure 9-10). The generic scope used in MultiSIM may not be as complex to look at as the scopes in the lab or on the test bench, but it is able to accomplish all that is necessary with MultiSIM circuits and problems. However, the Agilent oscilloscope (see Figure 9-11) simulated in MultiSIM represents an actual scope that you might use in industry; it is a true-to-life representation of that unit of test equipment. Notice that both figures display a 120 VAC signal as well as a 12 VDC signal.

2. Open circuit file **09-06a**. Notice that this generic scope is divided into four basic sections other than the **Display** itself; these sections are the functional

areas of the scope. The four sections are the **Timebase**, **Channel A**, **Channel B**, and the **Trigger** section. Each of these sections has a specific function to perform so that the oscilloscope is able to do its task—to display the

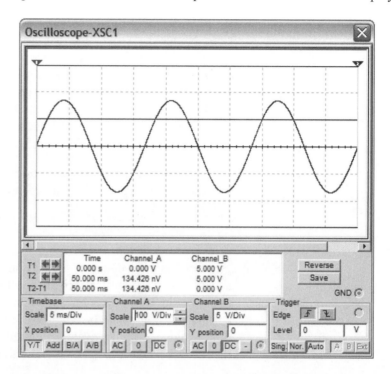

Figure 9-10   The Faceplate of the Generic MultiSIM Oscilloscope

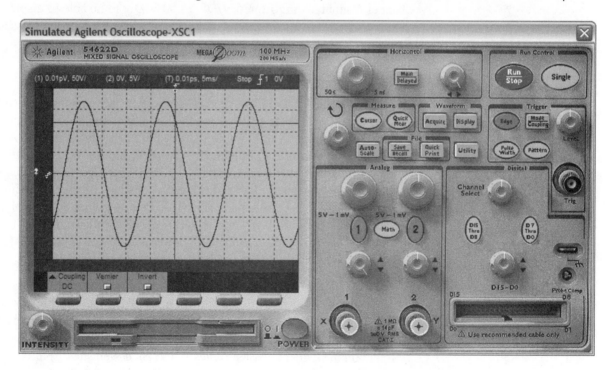

Figure 9-11   The Faceplate of the Agilent® Oscilloscope

operating characteristics of various types of electronics circuits. In this circuit, the scope is connected to a DC power source and is prepared to display a DC voltage.

3. Activate the circuit and observe the line that crosses the scope from left to right. Notice the X-axis line that crosses in the center of the scope from left to right. This line is calibrated to provide an indication of time relationships on the X-axis. The Channel A section connected to the DC power source is set for **5 V/Div**, which means that a large division on the vertical axis is equal to 5 volts. The signal line (trace) crosses the scope on the first division above the center, which means that the voltage being measured is equal to $1 \times 5\,V = 5\,V$. The X-axis in the center of the display represents the 0-V reference point.

4. Change the voltage of the DC power source to 10 V. Activate the circuit and observe where the DC line crosses the vertical axis. The line crosses the scope on the _____ large division line above 0 V, which is equal to

    _____.

5. Between the large divisions are five smaller divisions on the horizontal axis and spaces on the vertical axis. Each of these is equal to 20% of the value of one division. If one division were equal to 5 V on the vertical axis, then each smaller division would be equal to 1 V. Change the voltage output of the DC power source to 8 V. Activate the circuit and observe that the DC line crosses the vertical on the Y-axis about eight divisions above the center line.

6. Change the output of the DC power source to 100 V. You need to change the Channel A input setting from 5 V/Div to 50 V/Div to be able to display this larger voltage. Activate the circuit and observe where the DC line crosses the Y-axis again. The line crosses the Y-axis on the _____ division line, which is equal to _____.

7. Open circuit file **09-06b**. Notice that the Agilent scope is divided into a number of sections; these sections are the functional areas of the scope. The controls and sections that are important to us in our study of DC and AC are the **Display**, **Horizontal (Timebase)**, **Analog (Channel 1 and Channel 2 vertical input section)**, **Run Control**, the **Trigger** section, and (of course) the **Power** (on/off) switch. Each of these sections has a specific function to perform so that the oscilloscope is able to do its task, which is to display the operating characteristics of various types of electronics circuits. In this circuit, the scope is connected to a DC power source and is prepared to display a DC voltage. See Figure 9-12 for control and section locations on the oscilloscope.

    a. Activate the circuit and bring up the scope display (double left-click on the scope). This is going to be complicated; if you need additional help check out the 54622-97036.pdf file (the oscilloscope user's manual) that can be found on the Agilent or Electronics Workbench Web sites. Your computer has to have Adobe® Acrobat Reader installed to read PDF files.

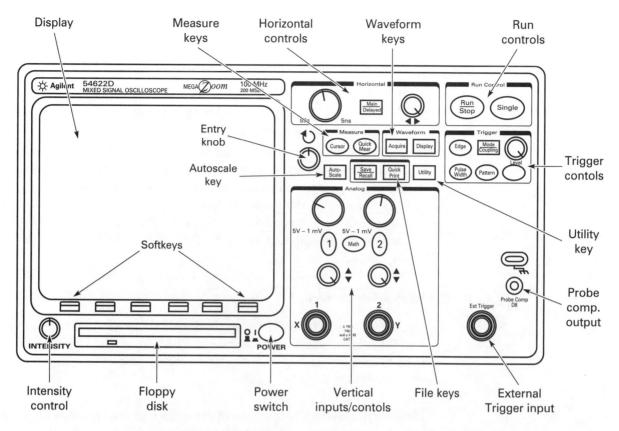

Figure 9-12 Agilent® 54622D Oscilloscope Front Panel

b. The power switch is one of the most obvious controls (lower right, below the display) on the scope. Turn it on.

c. Now, you need to adjust the Channel 1 input so that it can display the DC voltage. There are four large divisions above the X-axis; notice the white letters and numbers on the top left of the display, "**(1) 0V, 5V/.**" The 5 V indication means that the voltage value of each large division is 5 V. Use the **Volts/Division** knob (vernier) just above the **1,** as displayed in Figure 9-13, to change the input to **20V/Division**. You will see the top left display change from 5 V to 20 V as you click on the vernier.

Figure 9-13 Agilent Oscilloscope Input Voltage Control

d. At 20V/Division you should see the trace going across the o-scope. What is the voltage level? The trace is crossing the scope at _____ divisions above zero. The DC voltage level is _____.

e. You will gradually learn to use the Agilent oscilloscope as you progress through the following activities and chapters.

f. If you get the controls to a point where they are not responding as you expect or you just do not know how to progress any farther, you have two choices. One is to completely remove the scope from the desktop (highlight the scope and push the **Delete** key), and the other is to restore the scope to its default condition. This can be done by clicking on **Utility** in the **Waveform** section of the scope; then click on the default switch under the scope display.

● *Troubleshooting Problems:*

8. Open circuit file **09-06c**. The scope is already connected to the circuit under test. What is the value of the voltage at TPA? To see the DC line, you will need to change the Channel **A V/Div** setting. The voltage at the TPA is

_____.

9. If the horizontal line is off the top of the scope (the display), then the voltage is more than can be displayed by the Channel A V/Div setting. Increase the setting until the DC line (trace) is displayed. In this case, the proper V/Div setting is _____ V/Div.

10. Electromagnetic Devices

References:
Electronics Workbench®, MultiSIM Version 7
Electronics Workbench®, MultiSIM Version 7 User's Guide

Objectives   After completing this chapter the student should be able to:

- Use simple and variable inductor libraries in MultiSIM
- Use standard and nonlinear transformer libraries
- Use relays to remotely control circuits
- Connect DC motors and vary their output speed

Introduction

Electromagnetic devices are used throughout the electronics industry in component applications ranging from simple devices such as transformers, relays, and motors, to complex systems that depend on electromagnetic principles to attain desired effects. The principles of magnetism, electromagnetism, electromagnetic radiation, induction, and other related phenomenon have, over a period of several hundred years, infiltrated all aspects of modern industry and society.

The development of electrical power through the use of electromagnetic generators; the distribution of that electrical power through the capabilities of transformers; and the industrial application of that power, particularly with DC and AC motors, has brought about today's modern society. A return to the pre-electromagnetic days is almost impossible today. Third world countries around the globe are in catch-up mode as they attempt to achieve industrial and economic independence through modern industry and to develop the means by which that industry primarily operates today—the electric motor and AC electrical power.

In this section, the application of various electromagnetic devices found in the MultiSIM software will be discussed. These various types of electromagnetic devices represent only a small selection of the many components and systems available today. The technician needs to understand the theory as well as the applications of these magnetic, inductive, and electromagnetic devices. You will find at least some of these various applications in practically every piece of electronics equipment that you work with. Figure 10-1 on page 96 is a sampling of the many electromagnetic devices that can be found in and used in the MultiSIM software environment.

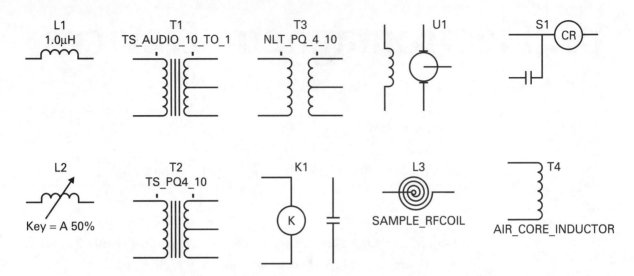

Figure 10-1 Electromagnetic Components Found in MultiSIM

Activity 10.1: Using the Inductors Found in MultiSIM

1. The **inductors** (coils) in MultiSIM are of two types: the simple inductor and the adjustable inductor. They are both found in the **Basic** menu. At this point, it is necessary to study inductors before attempting their application in electronic circuits. They are primarily used in AC circuits and do not react inductively in DC circuits. The only characteristic that it is necessary to be concerned about regarding inductive components in DC circuits is their DC resistance, the resistance of the wire that makes up the device.

2. Open circuit file **10-01**. These are some of the inductors (coils) that are found in MultiSIM (see Figure 10-2). Left double-click on the simple inductor (L_1) first and then open the **Value** tab. Notice that there is not much that you can do with this device except replace it. Left double-click on the virtual inductor (L_2), then on the **Value** tab where you will find that you can change the amount of inductance. The **Value** tab for the adjustable inductor (L_3) gives you the opportunity to change the key that changes inductance and the percentage of change. To change value, you have to replace the inductor from a menu selection. With the virtual adjustable inductor (L_4) you can change the inductance with the **Value** tab.

Activity 10.2: Using MultiSIM Transformers

1. There are two types of transformers used in MultiSIM: the standard transformer and the nonlinear transformer—both found in the **Basic** menu. Transformers are primarily used in AC circuits and do not normally operate in DC circuits. Transformers are much more complex than the MultiSIM components that have been studied up to this point. *Note:* For transformers to simulate in MultiSIM, both the primary and the secondary portions of the transformer circuit have to contain a ground. This is one case where

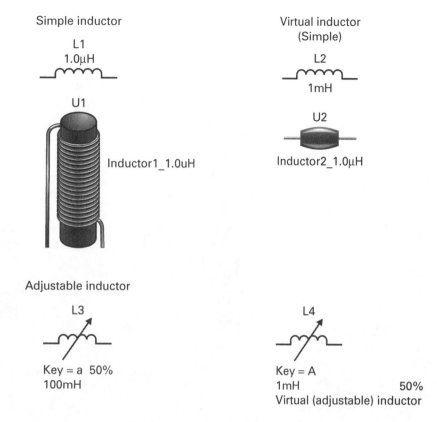

Simple inductor
L1
1.0µH

U1
Inductor1_1.0uH

Virtual inductor
(Simple)
L2
1mH

U2
Inductor2_1.0µH

Adjustable inductor
L3
Key = a 50%
100mH

L4
Key = A
1mH   50%
Virtual (adjustable) inductor

Figure 10-2   Inductors Used in MultiSIM

MultiSIM will not simulate in a manner that truly represents a real-world component where the secondary does not require a ground. Figure 10-3 on page 98 displays the menu choices for transformers from the **Basic** menu. Now that you are working in the AC realm of electronics, remember that panel meters have to be set to AC from the **Value** tab to operate correctly.

2. Open circuit file **10-02a**. The transformer that is displayed is a "standard" transformer selected from a menu and has a 10-to-1 step-down ratio. Use the DMM to measure the primary and secondary voltages.

   Measured primary voltage = _____.

   Measured secondary voltage = _____.

3. Does the transformer have the proper step-down ratio? Yes_____ or No_____.

4. Open circuit file **10-02b**. This circuit is the same as Step 2 except that the center tap on the transformer is grounded. The voltage, as measured from the center tap to the top and bottom pins of the secondary winding, should equal one-half of the secondary voltage previously measured.

   Measured center tap to top of the secondary voltage = _____.

   Measured center tap to bottom of the secondary voltage = _____.

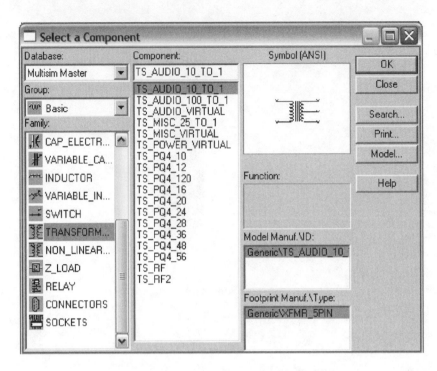

Figure 10-3 Transformer Library MultiSIM

- *Troubleshooting Problems*

5. Open circuit file **10-02c**. This circuit is not simulating properly; the secondary voltage should be 25 VAC, but U_1 displays 2.5 VAC. What is the problem?

The problem is _____

_____.

6. Open circuit file **10-02d**. This circuit is not simulating properly; the secondary voltage should be 25 VAC, but U_1 indicates that there is a problem with

the circuit. What is the problem? The problem is _____

_____.

Activity 10.3: Using MultiSIM Relays

1. Relays are electromagnetic switching devices with coils that are energized by current flow through the coil. The energizing action of the current flow causes a mechanical contact transfer to take place. There are many selections that can be made from the MultiSIM relay menu (see Figure 10-4). The basic differences between all of the models is the coil resistance. This is reflected in the amount of current necessary for the contacts to switch on (pull-in current, I_{on}) and the current that allows the contacts to switch off

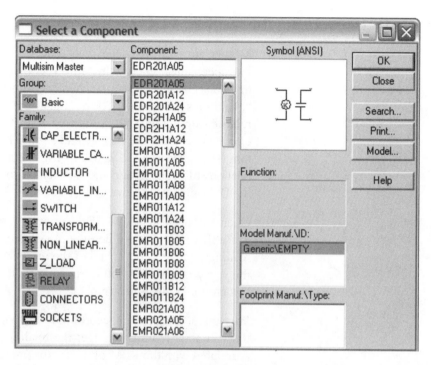

Figure 10-4 Basic Menu Showing the Relay

(drop-out current, I_{off}). To see these characteristics for a specific menu selection, left-click on **Model** (to the right when the menu is displayed) and the **Spice** simulation characteristics will be displayed. For example, the EDR201A05 unit being used in this activity has a coil resistance of 500 Ω and a pull-in current requirement of 7.5 mA. This means that the coil can be activated by a voltage source of 3.75 V.

2. Open circuit file **10-03**. This circuit is designed to operate a relay and control a separate circuit. When the circuit is activated and switch J_1 is closed, the relay switches on; its contacts close the controlled circuit containing the lamp. As a result of the switching action, the lamp, which is electrically isolated from the control circuit that energizes the relay, is turned on. Notice the switching action the relay contact when the switch is actuated. This is one of the primary uses of a relay, to isolate one power source from another. For example, in this circuit a DC power source in the relay coil circuit is isolated from the AC power source that provides power to the lamp. Activate the circuit and check for proper operation as indicated by lamp X_1 flashing and the closure of the contact (a diagonal line appears across the contact when it closes). Did the circuit operate properly? Yes_____ or No_____.

3. Install ammeter U_1 in the controlled circuit and measure the load current provided to the lamp.

   Load current = _____.

Activity 10.4: Using the MultiSIM DC Motor

1. Motors are rotating electromagnetic devices that change electrical energy into mechanical energy. In MultiSIM there is a virtual motor, a DC motor, that does not do anything mechanical. It just acts as a load and can provide an "evolutions per minute (RPM)" signal out to a meter to indicate its speed in relationship to the input voltage/current. This is an excellent tool to develop circuits to control DC motors and similar loads where output RPM is important. The speed of a DC in MultiSIM motor can be controlled by various connections of the armature and the field with a potentiometer and by changing the DC value of the power source.

2. Open circuit file **10-04**. This circuit demonstrates the operation of the MultiSIM DC motor. When the motor is energized, the voltmeter displays a reading relative to the speed of the motor. Varying armature current changes the speed of the DC motor. Speed variation and control is accomplished in this circuit by the potentiometer.

3. Vary the potentiometer according to the parameters of Table 10-1 and watch the voltmeter reading change; the reading on the voltmeter can be directly correlated with the RPM of the motor (not a real voltage output). Record the RPM data in Table 10-1.

	DC Motor Speed Correlated with Potentiometer Setting						
Potentiometer % Setting	0%	10%	25%	50%	75%	90%	100%
DC Motor RPM							

Table 10-1 DC Motor Speed Control

4. At what (percentage) setting of the potentiometer is it possible to achieve maximum RPM out of the DC motor?

   Maximum RPM is achieved at _____% potentiometer setting.

5. If it is desired to slow the motor down more than the minimum speed possible with the 10 Ω potentiometer, how can it be accomplished? In other words, what is the simplest method that can be used to slow this motor down more than has been attained with the present circuit? The easiest way to slow this motor down is by _____.

11. Alternating Voltage and Current

> **References:**
>
> *Electronics Workbench®, MultiSIM* Version 7
>
> *Electronics Workbench®, MultiSIM* Version 7 User's Guide

Objectives After completing this chapter the student should be able to:

- Use an oscilloscope to obtain voltage, frequency, and period (time) measurements of an alternating current (AC) waveshape
- Explain the relationships between frequency and period of an AC waveshape
- Discuss relationships between peak, peak-to-peak, RMS, average, and instantaneous values of AC waveshapes
- Understand phase relationships between waveshapes
- Harmonize relationships between AC waveshapes
- Determine characteristics of MultiSIM measuring instruments in AC circuits
- Analyze voltage, current, resistance, and power in resistive AC circuits

Introduction

Alternating current is used throughout the electrical and electronics industry in applications involving transformers, relays, and motors and also in complex systems that depend on electromagnetic principles to attain desired effects. The entire industrial base of modern society is based on the generation, distribution, and use of alternating current.

Direct current is unidirectional, generally fixed in value, and has polarity; alternating current varies in direction, value (amplitude), phase, and polarity. A key advantage of AC over DC is the ease in which voltage levels can be stepped-up or stepped-down by means of transformers—an easy change of voltage amplitude, which greatly benefits power distribution.

In MultiSIM there are three basic AC power sources (see Figure 11-1): the single-phase source (V_1), the Y-connected three-phase source, and the delta-connected three-phase source. AC signals can also be provided by test instruments such as function generators (found on the **Instruments** toolbar) or signal generators found on the test bench.

AC voltage readings on the meters are **RMS** (root-means-squared) readings, which can generally be correlated with working voltage. The outputs of the AC voltage and current sources and the function generators are also RMS signals. The oscilloscope will display peak-to-peak voltages. In the following activities, the usage of the oscilloscope will be studied extensively and developed as a valuable method that can be used to measure AC signals in electronics equipment.

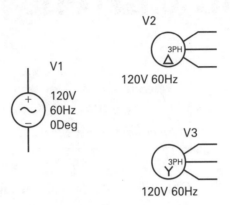

Figure 11-1 Basic AC Sources Found in MultiSIM

When displaying waveforms on the oscilloscope, click on the **Reverse** button on the scope control section and the background will change from black to white.

Activity 11.1: Various AC Waveshapes

1. The function generator produces three types of waveshapes: sinusoidal (the sine wave), triangular, and square.

2. Open circuit file **11-01**. First activate the circuit and then bring up the function generator and the oscilloscope (double left-click on them). In this circuit, an oscilloscope is connected to the output of the function generator. For the moment, the oscilloscope is to be used simply to observe the three types of waveshapes. Later, additional scope functions will be studied. Left-click on the waveshapes taskbar at the top of the function generator, one at a time, and observe the three types of displayed waveshapes displayed on the oscilloscope. The three types of waveshapes are _____, _____, and _____.

Activity 11.2: Measuring Time and Frequency of AC Waveshapes

1. In Chapter 9, the display of DC voltages on an oscilloscope was investigated. That investigation primarily consisted of learning about Y-axis displacement (vertical displacement) as a means of measuring DC voltage levels. In this chapter, interpretation of the various types of AC waveshapes displayed on the oscilloscope and an understanding of the meaning of these various types of scope data will be developed.

2. Open circuit file **11-02a**. The time relationship of the AC waveshape is displayed on the X-axis (horizontal displacement). In this case, the **Time-base** section of the scope is the area under consideration. This area of the scope is adjusted to determine various aspects of the AC waveshape in relationship to time. Observe that the time base is set for 200 μs/div, which means that one large division of horizontal deflection on the X-axis is equal to 200 microseconds (μs) in time. If a waveshape in which it took five divisions to complete a cycle of operation was displayed, then the time that was required to complete that cycle of the waveshape would be 200 microseconds × 5 divisions = 1.0 ms.

3. Activate the circuit and notice that the square wave displayed on the oscilloscope takes five divisions to complete one cycle of operation. For this type of waveshape, one cycle of operation is from the beginning of the waveshape on the left side of the scope face (where the trace goes up) to the end of one cycle of operation where the waveshape starts to go up again. The waveshape rises at the left, stays high (flat top) for 2.5 divisions, falls to a minimum point (below the X-axis), stays there for 2.5 divisions, and starts the cycle over again by rising once again. Notice that the waveshape rises two divisions above the X-axis, which is a voltage level of 10 V/Div × 2 divisions = +20 V. Then it falls to a point that is two divisions below the X-axis (which represents zero volts) for a voltage level of −10 V/Div × 2 divisions = −20 V. The waveshape has a top-to-bottom (peak-to-peak) value of 40 V, from −20 V up to +20 V. If necessary, the expanded view of the oscilloscope gives a larger view of the 0.1-ms divisions (see Figure 11-2). In this figure, a 1-kHz waveshape is being displayed on the generic MultiSIM oscilloscope as well as on the Agilent oscilloscope.

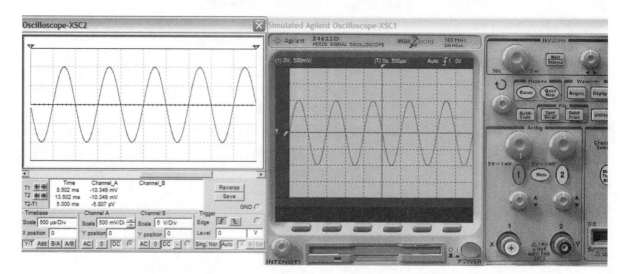

Figure 11-2 Oscilloscope Views of 1-kHz Sinewave

4. The formula **f = 1/t** develops the relationship between the time for one cycle of the displayed signal and how often a repetitive waveshape occurs per

second (frequency). The unit of measurement for **frequency** is the **hertz (Hz)**. If a waveshape occurs ten times per second, it occurs at a frequency of 10 Hz. In the case of the waveshape under observation in this activity, it takes 10×0.1 ms $= 1$ ms in time for one cycle and thus, its frequency is 1/0.001 or 1000 Hz (1 kHz).

5. Open circuit file **11-02b**. Activate the circuit and observe the signal on the MultiSIM scope face. Stop the waveshape movement by left-clicking on the pause button next to the on/off button above the display. Notice the red and blue vertical cursors that are positioned at the beginning and the end of the sine wave. Try moving them with the mouse. There is a window under the scope display (see Figure 11-3). The window has three lines of information. Line 1 displays Cursor 1 (T1) information (Time), the time into the sweep (500 μs from the left of the scope face) and the voltage (Channel_A) at the point where the cursor is located (where it crosses the signal display). Line 2 displays Cursor 2 (T2) information, which is the same kind of information as Cursor 1, time of 1.500 ms and a voltage. Line 3 displays time (T2–T1) and voltage differences between the two cursors. These cursors are valuable because they can be positioned on Channel A signals and measure time differences between two points as well as measuring voltage differences. If Channel B is displaying a signal along with Channel A, then Cursor 1 displays Channel A information in the window and Cursor 2 displays Channel B information in the window.

	Time	Channel_A	Channel_B
T1 ◄►	500.000μs	8.082nV	
T2 ◄►	1.500ms	54.145nV	
T2-T1	1.000ms	46.063nV	

Figure 11-3   The Cursor Information Window

6. In the case of this signal, place the cursors at the beginning and the end of a complete sine wave and window Line 3 displays the time difference. See Figure 11-4 for an example of cursor positioning.

    The time difference T2–T1 is approximately _____ and the frequency of

    the signal is _____ (f = 1/t).

●   *Troubleshooting Problem:*

7. Open circuit file **11-02c**. Activate the circuit and determine the time of the square wave displayed on the oscilloscope. Calculate the frequency of the square wave. The displayed time for one cycle is _____. The calculated frequency of the square wave is _____.

8. Notice that the generic square wave source is being used as a signal source; this is a DC signal. It never goes below the 0-V reference line (X-axis) in the middle of the Y-axis.

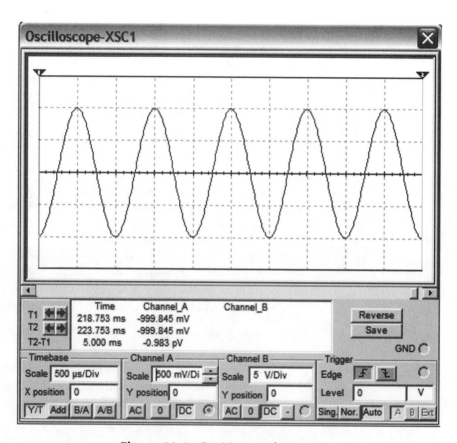

Figure 11-4   Positioning the Cursors

9. Open circuit file **11-02d**. Using this circuit, determine the frequency of the sine wave displayed on the oscilloscope. The time for one cycle is _____. The frequency of the sine wave is _____.

10. Open circuit file **11-02e**. In this circuit, change the oscilloscope time base setting (scale) to 20μs to be able to observe the period of one sine wave. Then determine the frequency of the sine wave displayed on the oscilloscope. The time for one cycle is _____. The frequency of the sine wave is _____.

Activity 11.3: Measuring Values of AC Waveshapes with the Oscilloscope

1. It is easy to determine the DC level of DC voltages on an oscilloscope because they are displayed as straight lines crossing the Y-axis at a determinable point. AC is harder to deal with because it is a continually changing waveshape and is very difficult to pin down to a specific value of voltage. There are other points of concern for the technician in the area of instrumentation. What are the voltage values, are they accurate, and are all of the instruments saying the same thing? In this activity, a better understanding of RMS, average, peak, peak-to-peak, and instantaneous values of waveshapes will be developed.

2. Open circuit file **11-03a**. Activate the circuit and observe the straight line going across the display. At the bottom of the Channel A section notice that the channel is set for **0** rather than AC or DC. When the channel is set for "0," the "zero" reference for the screen is displayed in the center of the screen on the X-axis (this can be adjusted by the offset arrows, moving the trace up and down as necessary). Click on the **DC** setting and observe the waveshape.

    This waveshape starts at _____ and rises to _____. Its frequency is _____.

3. Again, this is a DC waveshape of zero volts and it never goes below the zero reference line. *Because* it is zero volts, it never goes above the zero reference line either.

4. Open circuit file **11-03b**. Activate the circuit and stop the signal with the pause button. Position Cursor 1 at the highest point and Cursor 2 at the lowest point on the sinusoidal signal. Determine waveshape information from the Cursor Window. The Cursor Window indicates that the waveshape voltage rises to about _____ and that the waveshape falls to about _____.

5. The portion of the waveshape above the zero reference point is called the "**positive peak**" voltage and the waveshape below the zero reference is called the "**negative peak**" voltage. When referring to the overall waveshape from the most positive point to the most negative point, the term **peak-to-peak** is used. In this circuit, the peak positive voltage = +_____, the peak negative voltage = −_____, and the peak-to-peak voltage = _____.

6. Open circuit file **11-03c**. Activate the circuit and determine the peak and peak-to-peak values of the displayed signal. In this circuit, the peak voltages = +_____ or −_____ and the peak-to-peak voltage = _____.

7. **Instantaneous** voltage is the voltage at any specific point on the AC signal at that instant in time. When using the cursors in the expanded display, the T_1 and T_2 values displayed in the Cursor Window, in actuality, represent the instantaneous voltage value of the displayed signal wherever the cursor is located at that point on the Channel A or Channel B waveshape.

8. Open circuit file **11-03d**. Activate the circuit, expand the scope, and determine the instantaneous values as Cursor 1 crosses the first horizontal division above the center line and Cursor 2 crosses the first horizontal division below the center line. The instantaneous value for T_1 = _____ and for T_2 = _____.

9. To determine instantaneous values at a particular angular point on a sinusoidal waveshape, it is necessary to estimate the angular point on the scope display of the signal. If a more accurate measurement is desired, mathematical formulas will have to be used.

10. When dealing with RMS voltage values, the MultiSIM oscilloscope displays peak-to-peak signals. RMS values cannot be obtained without doing some calculations. RMS = 0.707 × peak voltage. So, a 100 V peak signal, as displayed on the oscilloscope, would equal 70.7 V_{RMS}. Most meters display RMS values.

11. Open circuit file **11-03e**. Activate the circuit and determine the peak, peak-to-peak, and RMS values. Peak voltage = _____, peak-to-peak voltage = _____, and RMS voltage = _____. You should have measured the peak and peak-to-peak voltages with the oscilloscope and the RMS voltage with the DMM.

12. When dealing with **Average** voltage values, it is necessary to go to mathematical calculations to determine the value. Average voltage = 0.637 × Peak voltage. So, a 100 V peak signal would equal 63.7 $V_{average}$.

Activity 11.4: Phase Relationships between Waveshapes

1. A point on an AC signal is often referred to in degrees of angular rotation, for example, the 30° point of a signal. When discussing **phase difference** between sine waves and **phase shift** of a signal in an electronics circuit, the term used to discuss the amount of phase difference or phase shift is also stated in degrees of angular rotation. The oscilloscope is an excellent tool to determine phase differences; you can use Channel A to observe one signal and Channel B to observe another signal and then compare their relationship in time. With signals, when you want to determine the amount of phase differences, the signals have to be of the same signal frequency, but can differ in voltage levels.

2. Open circuit file **11-04a**. In this circuit, there are two signal sources (V_1 and V_2). The wire from V_1 to Channel A on the oscilloscope is red and that signal will be red on the scope display. The wire from V_2 to Channel B on the oscilloscope is blue and that signal will be blue on the scope display. These colors will be standard operating procedure for the rest of this study: Channel A input will be red and Channel B input will be blue. Activate the circuit and observe that the two signals appear to be _____° apart in phase relationship.

3. Each value of phase shift, or phase difference, with most oscilloscopes is always an approximation. You determine, as closely as possible, the value in degrees of angular rotation for each division on the X-axis and approximate the phase difference from that knowledge. In this circuit, one sine wave is ten divisions in length, resulting in one division being equal to 36° in angular rotation. If there were four divisions between the two signals from the same relative point on the signal, then the amount of phase difference would be _____°.

4. Open circuit file **11-04b**. Once again, the value of one division on the X-axis is 36°. In this circuit, the phase difference is less than 180°. What is the approximate phase difference? The phase difference is _____°.

Activity 11.5: Harmonic Relationships Between AC Signals

1. A harmonic signal is a signal that is a multiple of a reference signal. For instance, the second harmonic of a 1-kHz signal would be 2 kHz. A third harmonic would be 3 kHz.

2. Open circuit file **11-05a**. Activate the circuit and observe that there are two blue sine waves for every red sine wave. Thus, the Channel B signal is a _____ harmonic of the Channel A signal.

3. Open circuit file **11-05b**. Activate the circuit and observe the signal differences between the two oscilloscope input channels. The Channel B signal is a _____ harmonic of the Channel A signal.

Activity 11.6: Characteristics of MultiSIM Measuring Instruments in AC Circuits

1. The voltage, current and power values in AC circuits are usually dealt with in terms of RMS values. Be careful to make sure that, when measurements are being made of values other than RMS values, there is not a mixture of values, such as mixing average, peak, and RMS voltage values. Any mixture of measurement values will cause an error in the calculations. The oscilloscope display has to be approached with caution because the displayed data is in the form of peak or peak-to-peak values and most measuring instruments other than the oscilloscope are RMS-reading instruments. As has been previously stated, when a signal is being measured with a voltmeter, the value is usually an RMS value. When measuring the same signal with an oscilloscope, the signal is a peak or peak-to-peak value.

2. The technician needs to know what the various signal source output voltages are in terms of RMS, peak-to-peak, peak, and/or average. The determination of what unit of measurement a specific MultiSIM instrument refers to is an important step in the troubleshooting process.

3. Open circuit file **11-06a**. The help menu for the multimeter states that the DMM is an RMS-reading instrument. It is connected to an AC signal source from the **Sources** menu. Activate the circuit and notice that the output voltage of the AC signal source is the same as the voltage reading on the DMM. At the same time, the oscilloscope displays a peak-to-peak waveshape. What is the value of the DMM in terms of peak voltage? The displayed DMM voltage can be calculated in terms of peak voltage to be _____.

4. The voltmeters and ammeters in the **Indicators** menu are very useful as in-circuit measurement devices and are frequently used throughout this text. These meters are also RMS-reading devices. Open circuit file **11-06b** and activate the circuit. Note the ammeter reading. U_2 displays a current value in RMS terms. The 1-kΩ resistor (according to Ohm's law) tells you the RMS value of the voltage source. Here is a common trick: put a 1-kΩ resistor in series with an ammeter to tell you what a voltage source value is. And, conversely, measure voltage across a 1-kΩ resistor and you will know the value of current flow through the resistor. Ammeter U_2 displays a current reading of

   _____ RMS/peak/peak-to-peak (circle the correct answer).

5. Connect U_1 to TPA and TPB. Reactivate the circuit and observe that there is something obviously wrong with voltmeter U_1. To solve the problem, left double-click on U_1, left-click on the **Value tab**, and change the meter **Mode** from **DC** to **AC**. Reactivate the circuit again and U_1 displays _____ V_{RMS} as it should.

6. Whenever it is necessary to use these generic panel meters to measure AC, make sure that the meter mode is changed from DC to AC.

- *Troubleshooting Problem:*

7. Open circuit file **11-06c**. Activate the circuit. There is a problem in this circuit. The current through U_1 is wrong according to circuit calculations. Determine the problem and correct it. The problem is _____

   _____.

Activity 11.7: Voltage, Current, Resistance, and Power in Resistive AC Circuits

1. The parameters of voltage, current, resistance, and power in AC resistive circuits are similar to the same parameters found in DC circuits regarding the relationships between Ohm's law, Kirchhoff's laws, and the power formulas.

2. Open circuit file **11-07a**. Activate the circuit and determine the value of the voltage drop across R_1 according to the oscilloscope display. Displayed signal = _____ peak-to-peak.

3. Connect the voltmeter to the circuit to verify that the two instruments are validating each other. V_{R1} = _____ $V_{RMS} \times 2.828$ = _____ peak-to-peak.

4. Do the two instruments validate each other? Yes _____ or No _____.

5. Open circuit file **11-07b**. Activate the circuit and determine the parameters of the circuit using the DMM, the oscilloscope, and your calculator. Record the data in Table 11-1.

	Voltage				Current			
	V_P	V_{PP}	V_{RMS}	V_{AVG}	I_P	I_{PP}	I_{RMS}	I_{AVG}
R_1								
R_2								
R_3								
Total								

Table 11-1   Circuit Parameters of a Resistive AC Circuit

● *Troubleshooting Problem:*

1. Open circuit file **11-07c**. Calculate circuit parameters and enter the data into Table 11-2. Then, activate the circuit and compare the calculations with the operating parameters of the circuit. There is a problem. What is it? The

problem is _____

_____ .

	Voltage				Current			
	V_P	V_{PP}	V_{RMS}	V_{AVG}	I_P	I_{PP}	I_{RMS}	I_{AVG}
R_1								
R_2								
R_3								
R_4								
R_5								
Total								

Table 11-2   Determining Circuit Parameters of a Resistive AC Circuit

12. Inductance and Inductive Reactance

> **References:**
> *Electronics Workbench®, MultiSIM* Version 7
> *Electronics Workbench®, MultiSIM* Version 7 User's Guide

Objectives After completing this chapter the student should be able to:
- Use a MultiSIM inductance meter to measure inductance of an inductor
- Calculate and measure inductance in series and parallel inductive circuits
- Calculate and measure inductive reactance and impedance in a circuit
- Determine phase relationships of current and voltage in an inductive circuit
- Calculate the Q and power consumption of an inductor
- Operate circuits containing inductance
- Troubleshoot inductive circuits

Introduction

Inductance (L) is one of the basic phenomena of electronics that is used in all types of electronics and industrial equipment throughout the modern world. There are many components and electronic circuits using inductance and inductive characteristics. Inductance is defined as the property in a circuit that causes opposition to a change in current flow and the ability to store a charge in an electromagnetic field. An **inductor** is a component that has inductance and may also be called a coil or a choke. The basis for an understanding of inductance is dependent upon an understanding of electromagnetic induction.

The definition of inductance—opposition to a change in current flow—tells us that inductance is an opposing element in alternating current circuits where there is a constant change of direction of current flow. In DC circuits, inductance has little effect except during the turn on and the turn off cycle or in the case of varying voltage levels such as pulsating DC or DC square waves.

Inductors have many uses in electronics. Typical uses for inductors as components include "tuned" circuits in radio and television circuits, as filters in electronic circuits to remove undesired signals and characteristics, in automotive applications such as the ignition coil, and so forth. Inductance is found in transformers, motors, relays, and many other electronic components.

Inductors also have the ability to store a charge. The charge is stored in the electromagnetic field that is built up around the component as a result of current flow. In a DC circuit, the electromagnetic field charge is initially built up around the inductor and remains there until current flow ceases. In an AC circuit, the electromagnetic field charge is constantly changing as the current continuously changes its direction of flow. Also, in an AC circuit, an inductor attempts to inhibit the change in current flow as the energizing force, the voltage, is continually changing direction. As the voltage tries to increase or decrease and as the current attempts to follow suit, the stored energy in the established electromagnetic field attempts to keep current from changing, thus opposing the change in current flow. The amount of opposition is dependent on the strength of the charge stored in the electromagnetic field.

The amount of opposition to the change in current flow that an inductive device offers to that attempted change in current flow is very important in an AC circuit. The term for the "amount of opposition" to a change in current flow is **Inductive Reactance**. Inductive reactance $(\mathbf{X_L})$ is measured in ohms in a manner similar to resistance. For example, $X_L = 10\ \Omega$.

The unit of inductance **(L)** is termed the "**Henry**" in honor of Joseph Henry, a pioneer in the field of electronics, and is abbreviated as **H**. For example, L = 10 mH.

Wire has resistance and coils are constructed of many turns of wire. In MultiSIM, the coils are referred to as "**ideal**" inductors and they have no resistance. MultiSIM inductors are found in the **Basic** library on the taskbar.

Activity 12.1: Measuring Inductance with an Inductance Test Circuit

1. In Chapter 1, the various types of inductive components and devices were introduced. In this activity, an inductive component, the coil, will be studied primarily under AC operating conditions. A coil is usually many turns of wire wrapped in a circular fashion around a coil form or core device.

2. Open circuit file **12-01a**. In this circuit a test circuit entitled "**Inductance Test Circuit**" is used to measure the inductance of a coil. This test circuit is designed using MultiSIM components to test unknown coils (see Figure 12-1). The voltmeter to the right is part of the test circuit and will read out the inductance of an unknown coil in units of inductance (Henrys). Ignore the "V" on the meter and substitute the symbol for Henrys, H, in its place.

3. Activate the circuit and determine the inductance of the unknown coil. The inductance of the unknown coil is _____.

4. Open circuit file **12-01b**. Determine the inductance of the three coils, L_1, L_2, and L_3 using the inductance test circuit. The measured inductance of the three coils is $L_1 =$ _____, $L_2 =$ _____, and $L_3 =$ _____.

Inductance Test Circuit

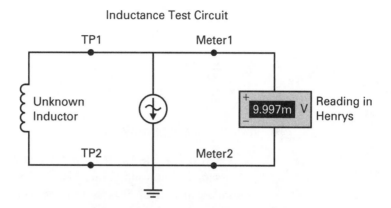

Figure 12-1 The Inductance Test Circuit

5. Open circuit file **12-01c**. In this circuit there is a 150 mH variable inductor. Determine the percentage (%) setting that provides 45 mH. The variable inductor is adjusted to a _____% setting to achieve a value of 45 mH.

● *Troubleshooting Problem:*

6. Open circuit file **12-01d**. In this circuit you are expecting 100 mA of current flow. Obviously there is a problem. Check the components against the schematic using the DMM and the inductance test circuit. The problem is

_____

_____.

Activity 12.2: Calculating and Measuring Inductance

1. Inductors in series and parallel are similar to resistors in series and parallel. When inductors are in series, the total amount of inductance in the circuit is equal to the sum of the individual inductance values. When inductors are in parallel, they use the same formulas as resistors in parallel to determine total inductance, except substitute inductance for resistance.

2. Open circuit file **12-02a**. Calculate the total inductance of this series circuit. The calculated total inductance of the circuit is _____.

3. Activate the circuit and measure the total inductance of the series coils. The measured total inductance of the circuit is _____.

4. Open circuit file **12-02b**. Calculate the total inductance of the parallel circuit. The calculated total inductance of the circuit is _____.

5. Activate the circuit and measure the total inductance of the parallel coils. The measured total inductance of the circuit is _____.

6. Open circuit file **12-02c**. Calculate the total inductance of the series-parallel circuit. The calculated total inductance of the circuit is _____.

7. Activate the circuit and measure the total inductance of the series-parallel coils. The measured total inductance of the circuit is _____.

- ● *Troubleshooting Problem:*

8. Open circuit file **12-02d**. This circuit has an unknown coil **(L₁)**, and it is necessary to know its value. Use the inductance test circuit to determine the inductance of L_1. It is best to completely disconnect the coil from the circuit path to avoid false measurements. Disconnect the coil and measure it. The measured inductance of L_1 = _____.

Activity 12.3: Calculating and Measuring Inductive Reactance in an AC Circuit

1. As stated previously, inductive reactance is the term used to describe opposition to change in current flow. The symbol for inductive reactance is X_L, and the unit of measurement is the ohm (Ω). There are two basic methods used to calculate inductive reactance. The first is the Ohm's law method determined in a manner similar to calculating DC circuits $(X_L = V/I)$ and the other method uses a formula that takes into account the operating frequency and the inductance of the inductor in the circuit. "Real" inductors have resistance (wire resistance) and inductive reactance. Inductive reactance is more commonly taken into consideration in an AC circuit rather than wire resistance. Remember that both parameters are there and sometimes both parameters are important. Figure 12-2 displays a circuit where Ohm's law can be used to calculate inductive reactance. Use Ohm's law to calculate X_L for this circuit.

    Calculated X_L $(X_L = V/I)$ for the circuit of Figure 12-2 = _____.

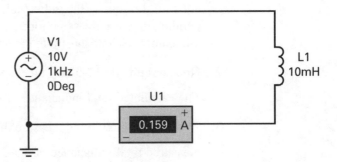

Figure 12-2   A Circuit Where Ohm's Law Can Be
Used to Calculate Inductive Reactance

2. Open circuit file **12-03a**. In this simple inductive circuit, the applied voltage is given and ammeter U_1 will display the amount of current that is flowing in

the circuit. Inductive reactance in this case can be solved by using Ohm's law, with one exception: X_L takes the place of R in the Ohm's law formula.

Activate the circuit and determine X_L. Calculated X_L = _____.

3. One big difference between resistive and inductive (reactive) circuits is the part the frequency of the AC signal plays in an inductive circuit, while having no effect in an operating DC circuit. The other formula you can use to determine X_L uses the inductance and frequency factors and is stated as $\mathbf{X_L = 2\pi fL}$. This formula has the constant $\mathbf{2\pi}$, $\mathbf{f}$ represents the frequency of the AC signal, and $\mathbf{L}$ represents the inductance of the inductive component(s). In any AC circuit, inductive reactance is directly related to the frequency of the signal source; if frequency is changed, then X_L will proportionally change.

4. Open circuit file **12-03b**. Using the formula $X_L = 2\pi fL$, calculate the inductive reactance of the circuit at the displayed frequency of 5 kHz. Then activate the circuit and use Ohm's law to verify the calculation. Calculated X_L =

   _____. Measured X_L = _____. Did they agree? Yes_____ or

   No_____.

5. Many times there will be slight differences between the measured and the calculated solutions to problems involving inductors and capacitors. The technician has to be able to determine when the discrepancies between measurements and calculations are important and when they are not.

6. Open circuit file **12-03c**. Using the formula $X_L = 2\pi fL$, determine the inductive reactance of this circuit at the displayed frequency of 1 kHz. Then activate the circuit and use Ohm's law to verify the calculation. Calculated X_L =

   _____. Measured X_L = _____.

7. Using the knowledge gained from these circuits, it is possible to make a statement concerning X_L, frequency, and inductance. When frequency goes up,

   X_L goes _____. When inductance goes up, X_L goes _____. When frequency goes down, X_L goes _____. When inductance goes down, X_L goes

   _____.

8. Open circuit file **12-03d**. Calculate the X_L ($X_L = 2\pi fL$) of this circuit and project the value of circuit current. Calculated X_L = _____. Calculated I_T

   = _____.

9. Activate the circuit and use Ohm's law to verify the calculations. Measured

   X_L = _____. Measured I_T = _____.

● *Troubleshooting Problems:*

10. Open circuit file **12-03e**. Calculate the X_L of this circuit and project the value

    of circuit current I_T. Calculated X_L = _____. Calculated I_T = _____.

11. Activate the circuit and verify the calculation concerning I_T. I_T seems to be incorrect according to the calculation. What is wrong? The problem with the circuit is _____.

12. Open circuit file **12-03f**. Calculate the X_L of this circuit and project the value of circuit current I_T. Calculated X_L = _____. Calculated I_T = _____.

13. Activate the circuit and verify the calculation concerning I_T. Again, I_T seems to be incorrect according to the calculation. What is wrong? The problem with the circuit is _____

_____.

Activity 12.4: Phase Relationships in Inductive Circuits

1. In a purely inductive circuit, current flow lags voltage by 90°. In other words, the sinusoidal voltage drop across an inductor will lead the current flow through it by 90°. In an electronics circuit with only an inductor in the circuit, it is difficult to display this relationship. It is easy to observe the voltage across the inductor with an oscilloscope (an oscilloscope can only display voltage), but it is not possible to observe the current through the inductor. The way around the problem is to insert a very small resistor in series with the inductor to monitor the circuit current (the current is the same through the resistor as through the inductor in a series circuit). See Figure 12-3 for a typical circuit prepared to measure phase shift. The small resistance will not affect the circuit to any great degree.

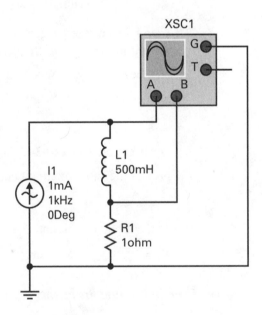

Figure 12-3   A Test Circuit to Measure
Phase Shift Through an Inductor

2. Open circuit file **12-04a**. In this circuit there is a 1-Ω resistor in series with the inductor. It is installed to monitor inductor current. The inductor opposes the current change in the sinusoidal signal applied to the circuit, and the resistor voltage drop reflects that opposition to a change in current flow and the resultant lag of the current behind the voltage. Activate the circuit and observe the 90° (almost) phase shift between the two quantities, the applied voltage and the inductor current (as seen by the voltage drop across the resistor). Stop the action of the scope traces with the **Pause** button. Channel A is displaying the voltage applied to the inductor, which is at a maximum point at the left of the screen. Channel B reflects the current flowing through the inductor, which is at zero at the left of the screen. This indicates that at the beginning of the sweep (of the scope traces), from left to right, voltage is at a maximum and current is at a minimum; this is a 90° phase difference. The 90° phase shift waveshapes are shown in the oscilloscope of Figure 12-4. The top waveshape displays the voltage applied to the circuit and the lower waveform displays the current flowing through the inductor. What is the inductive reactance of this circuit? Calculated X_L = _____.

3. The relationship between resistance and inductive reactance in a circuit influences the circuit in relation to the ratio of the X_L and R ohmic values. As seen in this circuit, X_L is much larger than R, so R does not have much influence.

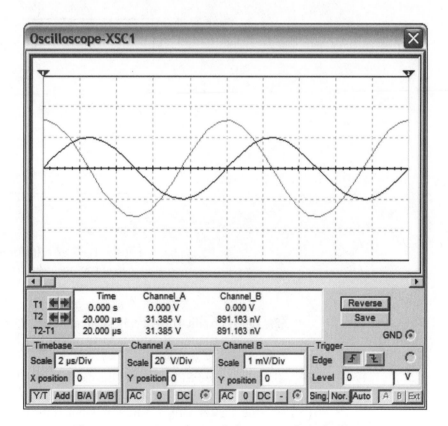

Figure 12-4   90° Phase Shift in an Inductive Circuit

4. Open circuit file **12-04b**. Activate the circuit. Is the phase shift 90°? Yes_____ or No_____.

5. Calculate the X_L of the circuit using the component values in the circuit. Calculated X_L =_____.

6. Insert ammeter U_1 into the circuit and recalculate X_L using Ohm's law. Recalculated X_L = _____.

Activity 12.5: The Q of an Inductor and Power Consumption in Inductances

1. The symbol for inductor quality is "**Q**." Real-world inductors not only inject inductance into a circuit, they also inject resistance, the resistance of the wire that makes up the coil plus some other resistive factors. The term for the resistance plus the other resistive factors is "effective series resistance" **(ESR)**. The formula for Q is $Q = X_L/ESR$. This formula indicates that Q is the ratio between the resistance and the reactance of an inductive circuit. An inductive circuit with an inductor and ESR is shown in the circuit of Figure 12-5.

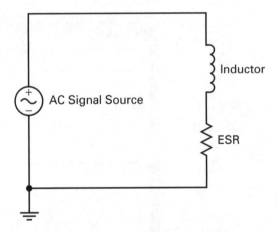

Figure 12-5 An Inductive Circuit Showing an Inductor with ESR

2. Open circuit file **12-05a**. Calculate the Q of this circuit. The first step to accomplish in this circuit is to calculate X_L. Calculated X_L = _____.

3. Using the ESR value and the calculated value for X_L, calculate Q. There is no unit of measurement for Q because it is a ratio of quantities of "like" units; the answer will just be a number (ratio). The calculated Q of the circuit is

_____.

4. An "ideal" inductor dissipates no power because it does not have resistance. A real-world inductor does dissipate power because there is always some resistance except in the case of superconductors. Power losses in a series inductive circuit are the results of ESR, which is the sum of copper loss (wire resistance), eddy-current loss, hysteresis loss, skin effect loss and dielectric loss. There are various techniques that are applied to inductors to reduce all of these losses with varying degrees of effectiveness.

5. What is the amount of power loss in this circuit? Use one of the power formulas to calculate the power loss. Calculated power loss = _____.

- **_Troubleshooting Problem:_**

6. Open circuit file **12-05b**. In this circuit, the current should be about 70.55 mA. Obviously there is something wrong. What is the problem? The problem is _____

   _____.

13. Resistive-Inductive Circuits

References:

Electronics Workbench®, MultiSIM Version 7

Electronics Workbench®, MultiSIM Version 7 User's Guide

Objectives   After completing this chapter the student should be able to:

- Operate series circuits containing resistors and inductors
- Operate parallel circuits containing resistors and inductors
- Operate series-parallel circuits containing resistors and inductors
- Troubleshoot resistive-inductive circuits

Introduction

A **resistive-inductive** (RL) circuit contains resistors and inductors in the form of components that exhibit those parameters. There is also stray resistance and stray inductance throughout a circuit. The electronic activity that takes place in an RL circuit is a combination of the characteristics of a resistive circuit and of an inductive circuit. Whether the circuit acts more resistive than inductive, or more inductive than resistive, depends on which parameter offers more opposition to current action than the other.

The approach to RL circuits is with some degree of "unknowing." On one hand there is a resistor, which has no voltage/current phase shift, in series with an inductor where the voltage leads the current by 90°. In the last chapter, low value resistors were used to aid in determining phase shift between voltage and current through an inductor; but these resistors did not affect the circuit parameters to any degree. In the case of large resistances in series with inductances, there will be considerable interaction.

It is very important to remember that current is the same throughout a series circuit. The current through the resistor is in-phase with the current through the inductor while the voltage dropped across the inductor is 90° out-of-phase with that current. It is not possible, under these circumstances, to simply sum up all of the individual voltage drops as is done in a simple DC circuit. Instead, a special method of addition called phasor addition will be used along with the Pythagorean theorem.

The opposition to current flow in an RL circuit consists of both resistance and inductive reactance and requires a modified approach to solving circuit parameters. The total opposition to current flow in a circuit containing reactive and resistive components, such as inductors and resistors, is called **Impedance (Z)**.

Activity 13.1: Series Resistive-Inductive (RL) Circuits

1. A series resistive-inductive (RL) circuit contains a resistor and an inductor in series as loads. Figure 13-1 is a series RL circuit as it is presented in MultiSIM. You can look at this circuit as resistive with an inductor added or inductive with a resistor added.

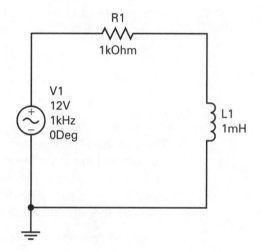

Figure 13-1   A Series Resistive-Inductive
Circuit in MultiSIM

2. Open circuit file **13-01a**. In this series RL circuit, an ammeter is installed to measure series current flowing in the circuit. Activate the circuit and determine current flow from the current meter. Measured I_T = _____.

3. Measure the voltage drops across the resistor and the inductor with the extra DMM (XMM2). Measured V_{R1} = _____ and V_{L1} = _____.

4. Does the sum of the voltage drops equal the voltage applied to the circuit by the voltage source (12 V)? Yes_____ or No_____.

5. As you can see, the sum of the voltage drops does not equal the voltage applied, as it should according to Kirchhoff's voltage law. The reason for this situation is that, while the current is the same through both of the series components, the resistor and the inductor, there is a phase difference (phase shift)

between the voltage drop across the resistor and the voltage drop across the inductor. The amount of phase shift depends upon the X-Y vector relationships between the voltage drops. There will be a 90° phase shift between the two voltage drops.

6. The solution to the "voltage drop" situation is found in phasor (addition) mathematics. Using the results of Step 3 and graph paper, draw a proportional phasor on the positive X-axis representing the voltage drop across the resistor. On the positive Y-axis, draw a proportional phasor representing the voltage drop across the inductor (this represents the 90° phase shift). And, finally, draw a phasor representing the hypotenuse of the two right angles on the plot. This hypotenuse can be measured to get a relative answer for total voltage (V_A), or the Pythagorean theorem method can be used to get a more exact answer. The answer using either method should equal the voltage applied to the circuit (12 V). Use the Pythagorean method to obtain an answer. The formula to use is:

Using the Pythagorean theorem formula, calculated V_A = _____.

$$V_A = \sqrt{VR^2 + V_L{}^2}$$

7. There are also two methods used to obtain a solution for the total opposition (impedance) to current flow problem: the Ohm's law method and the phasor mathematics method. The easiest is the Ohm's law method and it is usually used. In this circuit, determine impedance **(Z)** using Ohm's law. Z, in this case, replaces R in the Ohm's law formula **($Z = V_A/I_T$)**. Using the Ohm's law method of determination, Z = _____.

8. Now, using the Pythagorean theorem method, draw a proportional phasor on the positive portion of the X-axis on the plot representing the resistance of the resistor. Calculate the inductive reactance (X_L) of the inductor and then, on the positive portion of the Y-axis, draw a proportional phasor representing the inductive reactance of the inductor (this represents the 90° phase shift).

Calculated X_L = _____.

9. Finally, draw a phasor representing the hypotenuse of the two right angles on the plot. This hypotenuse can be measured to get a relative answer, or the Pythagorean theorem method can be used to get a more exact answer. This answer should be close to or equal the opposition to current flow that was previously calculated by Ohm's law. Calculate for Z using the Pythagorean theorem method. Calculated Z = _____.

10. Open circuit file **13-01b**. Determine circuit impedance (Z) and V_A. In this circuit, the first step is to calculate inductive reactance (X_L) and then impedance

(Z). Use the extra DMM (XMM2) to measure V_A. Calculated $X_L =$ _____. Calculated $Z =$ _____. Measured $V_A =$ _____.

11. Open circuit file **13-01c**. Determine X_L, Z, V_A, V_{R1}, V_L, and I_T. In this circuit, again, the first step to be taken is to solve for inductive reactance and then impedance. Use the DMM as much as possible. Calculated $X_L =$ _____ and $Z =$ _____.

12. Next, V_A can be determined by measuring the voltage drops across the - components and then by using the Pythagorean method of solution. Measured $V_R =$ _____ and $V_L =$ _____. Calculated (Pythagorean method) $V_A =$ _____.

13. Finally, I_T can be determined by Ohm's law. Calculated (Ohm's law) $I_T =$

_____.

14. Open circuit file **13-01d**. Determine X_L, Z, I_T, and V_A. $X_L =$ _____, $Z =$ _____, $I_T =$ _____, and $V_A =$ _____.

15. The determination of total power consumed in a series circuit also uses the phasor/Pythagorean method of calculation. The power consumed by the reactive (volt-amps reactive (VAR)) component and the resistive (watts) component is placed on each respective axis, and true power (V_A) is the hypotenuse of the right angles. $P_{R1} =$ _____W, $P_{L1} =$ _____VAR, and $P_T =$ _____V_A.

16. Open circuit file **13-01e**. Determine circuit parameters and enter the data in Table 13-1. Verify the R_1 calculation with the DMM. The measured value of $R_1 =$ _____.

	Current	Voltage	Power	Opposition to Current Flow
R_1			W	R =
L_1			VAR	X_L =
Totals			V_A	Z =

Table 13-1 Parameters of a Series RL Circuit

17. Open circuit file **13-01f**. Determine circuit parameters and fill in Table 13-2. Verify the R_1 calculation with the DMM. The measured value of R_1 = _____ .

	Current	**Voltage**	**Power**	**Opposition to Current Flow**
R_1			W	R =
L_1			VAR	X_L =
Totals			V_A	Z =

Table 13-2   Parameters of a Series RL Circuit

● *Troubleshooting Problems:*

18. Open circuit file **13-01g**. There is a problem with this circuit; the coil smoked and became history; also too much current is being drawn from the power source. Determine X_L and Z first, and then calculate the proper value of I_T. Calculated X_L = _____ , Z = _____ , and I_T = _____ .

19. Using the DMM, measure voltage drops and determine the fault. The problem is _____

_____

_____ .

20. Open circuit file **13-01h**. There is a problem with this circuit; the resistor is overheating and opened; also, too much current is being drawn from the power source. Determine X_L and Z first, and then calculate the proper value of I_T. Calculated X_L = _____ , Z = _____ , and I_T = _____ .

21. Using the DMM, measure voltage drops and determine the fault. The problem is _____

_____

_____ .

Activity 13.2: Parallel Resistive-Inductive (RL) Circuits

1. As we discovered in series DC (resistive) circuits, parallel RL circuits have identical voltage drops across every parallel component in the circuit (see Figure 13-2). The current will divide between the parallel branches according

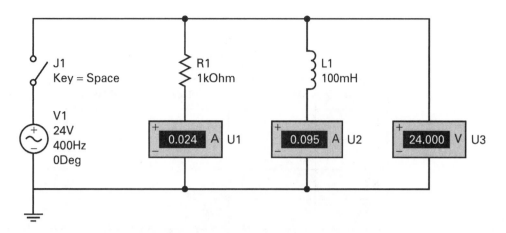

Figure 13-2   A Parallel RL Circuit in MultiSIM

to the amount of opposition (resistance or reactance) that each parallel branch offers to that flow of current. Phase relationships are another manner and differ from series RL circuits. In parallel RL circuits, the voltage is the same across all of the parallel components, but there is always a current phase shift between resistors and inductors in a parallel circuit. The current through inductors is shifted in-phase 90° from the resistors.

2. Open circuit file **13-02a**. This circuit displays the voltage drop (U_4) across the parallel components, the branch currents (U_2 and U_3), and the total circuit current flow (U_1). Activate the circuit and add the current flow through the parallel branches. Does the sum of the branch currents equal I_T as reflected by U_4? Yes_____ or No_____.

3. Does the voltage drop across the parallel branches as displayed by U_4 equal V_A? Yes_____ or No_____.

4. We will now use phasor addition to figure out why I_T does not equal the sum of the currents displayed by U_2 and U_3. Use the phasor method with resistor current reflected on the positive X-axis and inductive current reflected on the negative Y-axis (inductor current lags resistor current by 90°). Then use the Pythagorean method to determine circuit current. Calculated I_T =

_____.

5. The calculated current using phasors should equal the current displayed on meter U_1. Use Ohm's law and determine the reactance offered by the inductive branch of the circuit ($X_L = V_A/I_L$). Then determine the impedance using Ohm's law ($Z = V_A/I_T$). X_L = _____ and Z = _____.

6. Open circuit file **13-02b**. Determine the circuit parameters and place your results in Table 13-3. Remember that inductors in parallel act like resistors in parallel regarding total inductance.

	Current	Voltage	Power	Opposition to Current Flow
R_1			W	$R_1 =$
R_2			W	$R_2 =$
R_T			W	$R_T =$
L_1			VAR	$X_{L1} =$
L_2			VAR	$X_{L2} =$
L_T			VAR	$X_{LT} =$
Totals			V_A	$Z =$

Table 13-3 Parallel RL Circuit Parameters

7. In parallel RL circuits, the Ohm's law method is the easiest method to determine impedance when current and voltage parameters are available, but there is a mathematical method using one of the parallel resistance formulas and the Pythagorean theorem shown in Figure 13-3. What is Z using this formula? Calculated Z = _____.

$$Z = \frac{R_X X_L}{\sqrt{R^2 + X_L{}^2}}$$

Figure 13-3 Formula to Determine Impedance in a Parallel RL Circuit

● *Troubleshooting Problem:*

8. Open circuit file **13-02c**. Determine circuit parameters and fill in Table 13-4. The two resistors have the same value of resistance and should have the same amount of current flowing through them. Also, the two inductors, with the same values of inductance and inductive reactance, should have the same amount of current flowing through them. There appear to be two problems in this circuit. Find them. The two problems are _____

_____

and _____

_____.

	Current Should Be	Current Is	Voltage Is	Power Should Be	Opposition to Current Flow
R_1				W	$R_1 =$
R_2				W	$R_2 =$
R_T				W	$R_T =$
L_1				VAR	$X_{L1} =$
L_2				VAR	$X_{L2} =$
L_T				VAR	$X_{LT} =$
Totals				V_A	$Z =$

Table 13-4   Parallel RL circuit parameters

Activity 13.3: Series-Parallel Resistive-Inductive (RL) Circuits

1. In a manner similar to series-parallel DC circuits, series-parallel RL circuits are a combination of series sections where series-circuit parameters operate and parallel sections where parallel-circuit parameters operate (Figure 13-4). Each parallel RL section has to be simplified until it is minimized, hopefully,

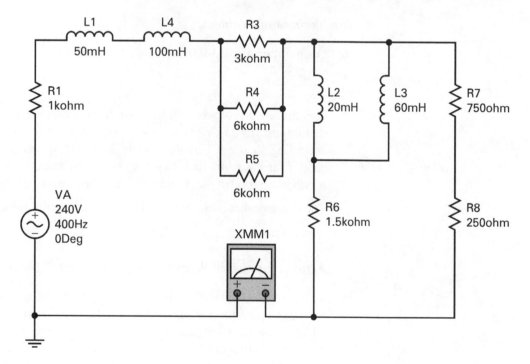

Figure 13-4   A Series-Parallel RL Circuit in MultiSIM

into a series RL section. Then further simplification can take place with series components. As with resistive circuits, the process of circuit simplification starts at the farthest point from the power source and proceeds toward the power source.

2. Open circuit file **13-03a**. Simplify this circuit. Activate the circuit and record the circuit current displayed by the DMM. Measured I_T = _____.

3. Combine resistors R_6 and R_7 into combination resistor R_{6-7} and resistors R_1, R_2, R_3, and R_4 into combination resistor $R_{1-2-3-4}$. Combine inductors L_1 and L_2 into combination inductor L_{1-2} and inductors L_3 and L_4 into combination inductor L_{3-4}. $R_{1-2-3-4}$ = _____, R_{6-7} = _____, L_{1-2} = _____, and L_{3-4} = _____.

4. Open circuit file **13-03b**. Change the value of the resistors and inductors to reflect the combination components of the previous circuit. Activate the circuit and record the circuit current. The current in this circuit, after changing the component values, should be the same as previously recorded in Step 2. Measured I_T = _____.

5. Open circuit file **13-03c**. Simplify this circuit into a single resistor and a single inductor. $R_{SIMPLIFIED}$ = _____ and $L_{SIMPLIFIED}$ = _____.

● *Troubleshooting Problems:*

6. Open circuit file **13-03d**. Activate the circuit and record I_T. Measured I_T = _____.

7. The circuit current seems to be too high; it is supposed to be about 29 mA. In this circuit, L_1 is defective, and it is supposed to be a 122 mH inductor according to the schematic. While looking through the parts bins in the shop, four inductors that appear to be of the same type as L_1 were located. Measure all four of them and replace L_1 with the one that is closest in value.

    The inductor that is closest in value to the schematic value of L_1 is _____; it measures _____ in inductance.

8. Activate the circuit. What does I_T measure now? Measured I_T = _____.

9. The part was replaced and the equipment turned on. After the equipment has been operating for some time, smoke rises from the equipment and the replacement for L_1 is history. Upon continuing the troubleshooting process, another bad component is found that is causing L_1 to smoke and it is replaced. The equipment has to be operational while parts are on order. Is there any

series-parallel combination with the other three coils that could temporarily replace L_1 until the correct replacement part arrives? If so, what is the combination? A possible solution would be _____

_____ and the series-parallel combination would measure _____ in inductance.

10. Change the parts and activate the circuit again. Check the value of I_T, which should be close to the measured value of Step 8. Measured I_T = _____.

Activity 13.4: Pulse Response and L/R Time Constants of RL Circuits

1. So far in our study of RL circuits, circuit operation under sinusoidal conditions has been emphasized. Now we are going to look at the action of RL circuits under non-sinusoidal conditions such as with square waves and the unique situation in DC circuits at turn-on and turn-off time. The operation of an RL circuit under squarewave conditions is similar to the turn-on and turn-off situation in a DC circuit where there is a sudden transition from off-to-on or on-to-off. Inductive components normally react to these sudden changes that take place in a DC circuit, but after that initial reaction, cease to respond to normal DC operating conditions.

2. Open circuit file **13-04a1**. In this circuit, we are going to view the increase in V_A from 0 V to 10 V over several time constants with the scope connected across the resistor. Channel A of the oscilloscope is connected across the resistor (R_1).

3. Activate the circuit to watch the rise in voltage from 0 V to 10 V in 5 TC. Determine the voltage two seconds after the point when the rise time starts.

The voltage after two seconds = _____.

4. You may place the red or blue cursor on the two-second point and read the voltage level directly from the related window.

5. Next, open circuit file **13-04a2**. This circuit and some of the following circuits have additional components (R_2 and D_1); do not be concerned with them, they are in the circuit to make the oscilloscope display give you an accurate picture of what is happening in the circuit.

6. Activate the circuit and view the decrease in V_L from 10 V to 0 V over 5 TC. Notice that the oscilloscope is set so that one major division is equal to 1.0 second. What is the duration of one time constant? 1 TC = _____ seconds.

7. Once again, the oscilloscope had proven to be a powerful tool to view the inner workings of an electronic circuit. If the V_R and V_L waveshapes were displayed simultaneously, the oscilloscope would display the two curves crossing each other in a manner similar to the Universal time-constant chart. What is the time duration of five time constants in this circuit? 5 TC = _____seconds.

8. What is the voltage after five time constants? The voltage after five time constants is _____.

9. Open circuit file **13-04b1**. Determine the voltage rise in time constant intervals and enter the data in Table 13-5.

Time Constant	1 TC	2 TC	3 TC	4 TC	5 TC
Voltage Reading					

Table 13-5 Voltage Rise Time in Time Constant Intervals

10. Open circuit file **13-04b2**. Determine the voltage fall time in time constant intervals and enter the data in Table 13-6.

Time Constant	1 TC	2 TC	3 TC	4 TC	5 TC
Voltage Reading					

Table 13-6 Voltage Fall Time in Time Constant Intervals

11. An RL circuit has a predicable rise and fall time when sudden transitions occur in a DC circuit. The response of an RL circuit, or any circuit, to the fast rise and fall time of square waves is called "step response."

12. Open circuit file **13-04c**. Activate the circuit and observe the step response of an RL circuit to a squarewave input. The response across the inductor is being compared with the input square wave. The upper red trace on the oscilloscope is the squarewave input and the lower blue trace is the **differentiated** response of the inductor (the A input is offset to facilitate comparison). This is considered to be a short time constant circuit. When the step occurs in this circuit, the inductor initially drops all of the voltage and then, after a period of five time constants, assumes its normal DC response of 0 V drop.

What is the duration of five time constants in this circuit? 5 TC = _____.

13. Open circuit file **13-04d**. Activate the circuit and observe the step response in two modes of **integrator** action. The display for the short time constant mode

is the type of signal that is usually considered to be an output of an integrator circuit. The display for the long time constant circuit has some integrator action as is observable from the rounding off of the leading and lagging edges of the square wave. The only difference between the two time constant circuits is in the size of the resistor. If the resistance value of R_2 were made larger, then the integration action would be less observable. What is the duration of the long and the short time constants in this circuit? The long time constant = _____ and the short time constant = _____.

● *Troubleshooting Problems:*

14. Open circuit file **13-04e**. You are developing this circuit for the engineering department and the circuit operating specification is for an extremely short time constant resulting in a differentiated pulse that looks like a spike as displayed in Figure 13-5.

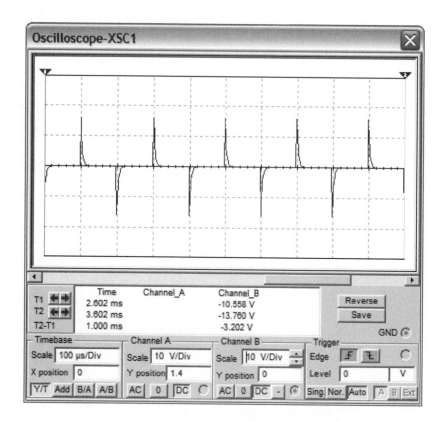

Figure 13.5 A Differentiated "Spike" Waveshape

15. Activate the circuit and observe the signal on the oscilloscope. There are two choices in this circuit: (1) to increase/decrease the resistance of the resistor, or (2) to increase/decrease the inductance of the inductor. Which would be the least expensive? Probably to _____ would be less expensive.

16. Try changing the component values until you get a signal like the one shown in Figure 13-5. What component values did you end up with to satisfy the engineering specifications? The values R_1 = _____ and L_1 = _____ satisfied circuit requirements.

14. Capacitance and Capacitive Reactance

References

Electronics Workbench®, MultiSIM Version 7

Electronics Workbench®, MultiSIM Version 7 User's Guide

Objectives After completing this chapter the student should be able to:

- Use an oscilloscope and a DMM to check a capacitor
- Calculate and measure capacitance in series, parallel, and series-parallel capacitive circuits
- Calculate and measure capacitive reactance and impedance in a circuit
- Determine phase relationships of current and voltage in a capacitive circuit
- Operate circuits containing capacitance
- Troubleshoot capacitive circuits

Introduction

Capacitance (C) is another of the basic phenomena of electronics and is used in many applications. There are many industrial applications that use capacitance and capacitive characteristics to their advantage. Capacitance is defined as the property, in a circuit, that opposes a change in voltage and has the ability to store a charge in an electrostatic field. A **capacitor** is a component that has capacitance and is also, historically, called a condenser. The basis understanding capacitance is closely related to an understanding of electrostatic charges.

Capacitors have many uses in electronics including being used in the "tuned" circuits found in the fields of radio and television transmission and reception, and as filters in electronic circuits to remove undesired signals and characteristics.

The first property of capacitance, opposition to a change in voltage, indicates that capacitance is an active device in alternating current circuits where there is a constant change of source voltage and direction of current flow. In DC applications, capacitors block current flow through them and thereby attempt to keep voltage constant in DC circuits where there is a change in voltage levels.

The second characteristic of a capacitor is its ability to store a charge in an electrostatic field. The charge is stored in an electrostatic field built up on the conducting surfaces that constitute the capacitor. In a DC circuit, the electrostatic field is built up on the conducting surfaces and remains there until power is removed or there is a change in the DC voltage level. In an AC circuit, the electrostatic charge of the capacitor is constantly changing as the source voltage changes. This change in source voltage and capacitor charge is reflected in circuit current flow. As the current tries to change direction and the voltage tries to follow suit, the stored energy in the established electrostatic field attempts to keep the voltage from changing, thus opposing the change in voltage level.

Another concept that is necessary to know concerning capacitance and capacitors is "how much opposition do they offer to a change in voltage levels." The term for the "amount of opposition" to a change in voltage level is "**Capacitive Reactance**." Capacitive reactance (X_C) is measured in ohms the same as resistance and inductive reactance. For example, $X_C = 10 \ \Omega$.

In MultiSIM, the capacitors are found in the **Basic** library on the taskbar. There are many types of capacitor choices offered in the basic components in this library and they are all considered to be "**ideal**" components, having the exact value assigned to them. Capacitance meters are commonly available in the electronics lab and are necessary to measure capacitance in and out of circuit during the troubleshooting of electronics equipment. In the MultiSIM environment, we will observe the charging action of a capacitor on an oscilloscope and a DMM as a means of determining capacitor action as well as a way to isolate defective capacitors when troubleshooting.

Activity 14.1: Checking Capacitance with an Oscilloscope and a Meter

1. In Chapter 1, various types of capacitors from the MultiSIM component library were introduced. Now we will study capacitors and the phenomenon of capacitance and capacitive reactance. In its simplest terms, a capacitor consists of two conducting surfaces separated by an insulator (a dielectric). In this activity we will look at the use of the oscilloscope and the DMM to check capacitors by observing their charging characteristics.

2. Open circuit file **14-01a**. In this circuit, an oscilloscope and a DMM are connected to test the functionality of a capacitor. This simple testing circuit is displayed in Figure 14-1. Activate the circuit and observe the action of the charging waveshape on the oscilloscope and the DMM. Toggle the switch and observe the rise and fall time of the waveshape. Use the oscilloscope to determine approximate rise and fall times of the charging waveshape.

3. The charging waveshape takes _____ seconds to charge to 10 V and falls in less than _____ second(s). The fall time of the waveshape is dependent on the value of the discharge resistor (R_2). If you changed the value of R_2 to the same value as R_1, the discharge waveshape would have the same duration as the charging waveshape, but in an opposite direction. Try it; change the value of R_2. After changing the value of R_2 to _____, the discharging waveform takes _____ seconds to fall to 0 V.

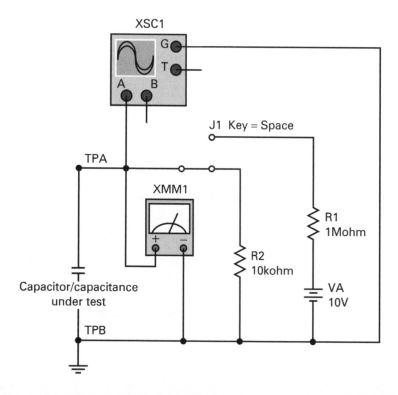

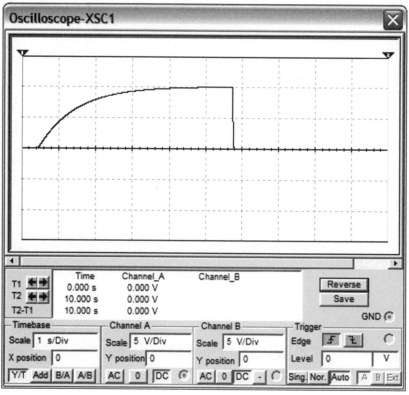

FIGURE 14-1   A Capacitor Charging Circuit

4. Open circuit file **14-01b**. Determine which of these capacitors (C_1, C_2, C_3, and C_4) are functional and which one is defective by connecting them into the circuit. Capacitors C___, C___, and C___ are functional. C___ has a problem.

5. Open circuit file **14-01c**. In this circuit there is a variable capacitor. This variable capacitor has to be adjusted to a setting of _____% to achieve a value of 3.0 μF.

● *Troubleshooting Problems:*

6. One simple test of capacitors is by using an ohmmeter to see if they will charge. This is a pretty rough test, giving little information except that the capacitor seems to be working properly. It does not give a capacitor value; it simply tells the technician whether the capacitor will charge, is a short, is open, or is leaky. With experience or with "known-to-be-good" comparison capacitors, technicians can determine whether a capacitor is charging correctly.

7. Open circuit file **14-01d**. In this circuit there are six capacitors that have to be tested to determine their condition. Four appear to be good, one is shorted (0.0000 Ω) and one is leaky (100 Ω). Identify the conditions and drag the correct title over from the right side of the screen and place it under the correct capacitor. C_1 is _____, C_2 is _____, C_3 is _____, C_4 is _____, C_5 is _____, and C_6 is _____.

8. Actually, one of the four capacitors that appear to be good is open, and we will have to perform an additional test to determine which one is defective.

9. Open circuit file **14-01e**. These are four of the capacitors that you tested in the previous exercise. They all tested good using the DMM as an ohmmeter to verify that they were not shorted or leaky. Now, you will check their charging capabilities using the DMM again, but this time as a voltmeter.

10. Activate the circuit and check each capacitor with the test circuit. Three of these capacitors are good and one is open. C_1 is _____, C_2 is _____, C_3 is _____, and C_5 is _____.

Activity 14.2: Calculating and Measuring Total Capacitance

1. The method of calculation used to determine the total capacitance for capacitors in series is performed in a manner similar to resistors and inductors in parallel, using one of the parallel formulas such as: $C_1 \times C_2 / C_1 + C_2$, or one of the other parallel (resistor) formulas.

2. The determination of total capacitance for capacitors in parallel is conducted in a manner similar to resistors and inductors in series; the capacitance values add and total capacitance is equal to the sum of all the capacitor capacitance values. Various circuit configurations are shown in Figure 14-2.

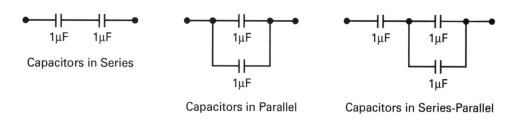

Capacitors in Series   Capacitors in Parallel   Capacitors in Series-Parallel

FIGURE 14-2   Examples of Capacitive Circuit Configurations

3. Open circuit file **14-02a**. Calculate the total capacitance of this series circuit. The calculated total capacitance of the circuit is _____.

4. Open circuit file **14-02b**. Calculate the total capacitance of this parallel circuit. The calculated total capacitance of the circuit is _____.

5. Open circuit file **14-02c**. Calculate the total capacitance of this series-parallel circuit. The calculated total capacitance of the circuit is _____.

● *Troubleshooting Problem:*

6. Open circuit file **14-02d**. There is a problem in this circuit; the current flow is supposed to be 4.835 mA and it is 5.488 mA. Use the second DMM to measure the components out of circuit. The problem in the circuit is

_____

_____.

Activity 14.3: Calculating and Measuring Capacitive Reactance in an AC Circuit

1. As previously stated, capacitive reactance (X_C) is the term used to delineate the amount of opposition to a change in voltage offered by a capacitor with that opposition being contained in the electrostatic charge on the capacitor. The unit of measurement for capacitive reactance is the ohm. Capacitive reactance can be calculated and its effects measured in a circuit using Ohm's law and Kirchhoff's laws in a manner similar to resistive and inductive circuits. "Real" capacitors not only have capacitive reactance, but they also have some power losses due to the resistance of the conducting surfaces, leakage resistance, and dielectric dissipation. We will ignore these power

losses in this study. MultiSIM capacitors are to be considered "ideal" with no power losses. Capacitive reactance is the most important factor regarding capacitors that needs to be considered in a circuit, but remember that the power loss problems are always there too.

2. The easiest way to calculate capacitive reactance in a series circuit is the Ohm's law method: divide the circuit voltage by the circuit current. See Figure 14-3, a circuit demonstrating where Ohm's law parameters can be used to calculate capacitive reactance.

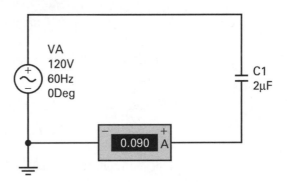

FIGURE 14-3   Using Ohm's Law to
Calculate Capacitive Reactance

3. Open circuit file **14-03a**. In this simple capacitive circuit the applied voltage is given and the ammeter displays the amount of current that is flowing in the circuit. Activate the circuit and use Ohm's law to solve for capacitive reactance. Note that X_C takes the place of R ($X_C = V_A/I_T$). Activate the circuit and determine X_C. Measured X_C = _____.

4. One big difference between resistive and capacitive (reactive) circuits is the part that the frequency of the AC signal plays in a reactive circuit while having little or no effect in a DC circuit. The formula for X_C, (using the capacitance factor), is $X_C = 1/2\pi fC$. In this formula, 2π is a constant (6.28), f stands for the signal frequency that the capacitor is responding to, and C stands for the amount of capacitance of the capacitive component (most commonly in μF). In any AC circuit, capacitive reactance has an exact value at a certain frequency of the signal source, and if the frequency is changed, then X_C will change proportionally.

5. Open circuit file **14-03b**. Using the X_C formula, determine the capacitive reactance of the circuit at the given frequency of 500 Hz. Calculated X_C = _____.

6. Activate the circuit and use Ohm's law to verify the calculation. Measured X_C = _____.

7. Open circuit file **14-03c**. Using the X_C formula, determine the capacitive reactance of the circuit. Notice that in this circuit the capacitance is less than 1μF. Calculated $X_C =$ _____.

8. Activate the circuit and use Ohm's law to verify the calculation. Measured $X_C =$ _____.

9. Using the knowledge gained from the circuits that you have just worked with, make a statement concerning X_C and frequency. When frequency goes up, X_C goes _____. When frequency goes down, X_C goes _____.

10. Open circuit file **14-03d**. Calculate the X_C of this circuit and project the value of circuit current. Calculated $X_C =$ _____. Projected $I_T =$ _____.

11. Activate the circuit and use Ohm's law to verify the calculations. Measured $I_T =$ _____. Calculated $X_C =$ _____ (Ohm's law method).

● *Troubleshooting Problems:*

12. Open circuit file **14-03e**. Calculate the X_C of this circuit and project the value of circuit current flow. Calculated $X_C =$ _____. Calculated $I_T =$ _____.

13. Activate the circuit and verify the calculation concerning I_T. Obviously there is a problem; I_T is not correct according to calculations. Use the DMM to determine the problem. The problem with the circuit is _____

_____

_____.

Activity 14.4: Phase Relationships in Capacitive Circuits

1. In a purely capacitive circuit, voltage lags current flow by 90°. In other words, the sinusoidal charging current flow onto the capacitor plates will lead the voltage drop across the capacitor by 90°. In an electronics circuit with only a capacitor in the circuit, it is difficult to display this relationship. It is easy to observe the voltage across the capacitor with an oscilloscope (an oscilloscope can only display voltage), but it is not possible to observe the current flow to the capacitor. The way to get around the problem is to insert a very small resistor in series with the capacitor to monitor the circuit current (the current is the same through the resistor and the capacitor in a

series circuit). Figure 14-4 displays a typical circuit to measure phase shift. This solution is similar to the method used with inductors to determine voltage phase shift.

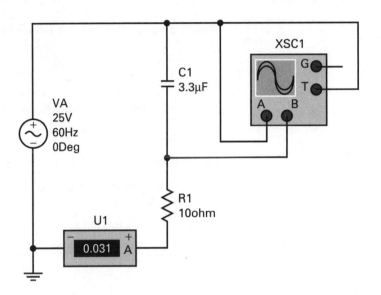

FIGURE 14-4   Circuit to Measure Phase Shift
in a Capacitive Circuit

2. Open circuit file **14-04a**. In this circuit the 10 Ω resistor in series with the capacitor is used to monitor capacitor current. The capacitor opposes the voltage change in the sinusoidal signal applied to the circuit and the resistor voltage drop reflects the AC voltage and its resultant lag behind the current.

3. Activate the circuit and observe the 90° (almost) phase shift between the two inputs, source voltage (V_A) and circuit current (I_T). Channel A is displaying the voltage applied to the capacitor, which starts at zero at the left of the screen. Channel B reflects the current flowing through the resistor, which is high at the left of the screen when the voltage is at zero. This indicates that at the beginning of the sweep (of the scope traces), from left to right, voltage is at a minimum and current is at a maximum; this is a 90° phase difference. Figure 14-5 shows the phase shift with the larger waveshape displaying the voltage applied to the circuit and the smaller waveshape displaying the current flowing to the capacitor.

4. Open circuit file **14-04b**. Activate the circuit. Is the phase shift 90°? Yes_____ or No_____.

5. What is the X_C of the circuit? Calculated $X_C =$ _____.

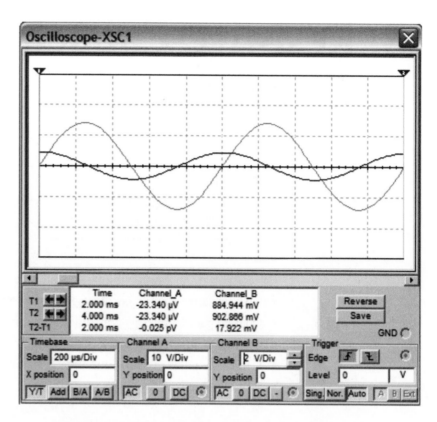

FIGURE 14-5   90° Phase Shift in a Capacitive Circuit

● *Troubleshooting Problems:*

6. Open circuit file **14-04c**. Activate the circuit. Is the phase shift 90°? Yes_____ or No_____. How many degrees (approximately) is the phase shift? The phase shift is approximately _____°.

7. What should the X_C of the circuit be? Calculated X_C = _____.

8. What is the fault in the circuit? The fault is _____.

15. Resistive-Capacitive Circuits

References

Electronics Workbench®, MultiSIM Version 7

Electronics Workbench®, MultiSIM Version 7 User's Guide

Objectives After completing this chapter the student should be able to:

- Operate series circuits containing resistors and capacitors
- Operate parallel circuits containing resistors and capacitors
- Operate series-parallel circuits containing resistors and capacitors
- Determine pulse response and time constants of RC circuits
- Troubleshoot resistive-capacitive circuits

Introduction

A **resistive-capacitive (RC)** circuit contains resistance and capacitance usually in the form of components that exhibit those parameters. The activity that takes place in a circuit of this type is a combination of both characteristics: the interaction of a resistive circuit with that of a capacitive circuit. Whether the circuit acts more resistive than capacitive, or more capacitive than resistive, depends on which parameter offers more opposition to current flow than the other.

When looking at RC circuits, observe that this type of circuit is a combination of resistors and capacitors where there is no voltage and current phase shift concerning the resistor, yet there is a 90° voltage versus current phase shift concerning the capacitor. It is not possible, under these circumstances, to sum up the individual voltage drops as it is done in a simple DC circuit. Instead, the same mathematical tools of phasor addition along with the Pythagorean theorem that were developed in the chapters on inductance will be used.

Activity 15.1: Series Resistive-Capacitive (RC) Circuits

1. A series resistive-capacitive (RC) circuit contains, as a minimum, a resistor and a capacitor in series as loads. In Figure 15-1 a series RC circuit as it is presented in MultiSIM is displayed. This series circuit contains an AC signal source, a capacitor, and a resistor.

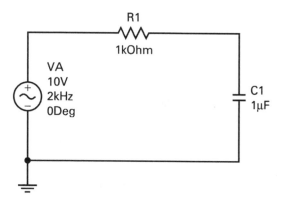

Figure 15-1   A Series RC Capacitive Circuit

2. Open circuit file **15-01a**. In this series RC circuit, an ammeter is installed to measure series current flowing in the circuit. Remember, according to Ohm's law, current is the same everywhere throughout a series circuit. Also, resistive and reactive components drop voltage in proportion to the amount of their resistance or reactance after taking into consideration the 90° phase shift.

3. Activate the circuit and determine current flow. Measured I_T = _____.

4. Measure the voltage drop across both the resistor and the capacitor using the second DMM. Measured V_{R1} = _____ and V_{C1} = _____.

5. Does the sum of the voltage drops ($V_{R1} + V_{C1}$) equal the voltage applied to the circuit by the voltage source (V_A)? The apparent sum of the voltage drops _____ (does/does not) equal the voltage applied. Does your answer agree with the precepts of Kirchhoff's voltage law for series circuits? Yes_____ or No_____.

6. There is a problem. The primary reason for this problem lies in the fact that, while the current is the same through both of the series components, the resistor and the capacitor, there is a phase difference between the voltage drop across the resistor and the voltage drop across the capacitor. The amount of this difference (phase shift) depends upon the vector relationship between the resistance and the capacitive reactance: both components impede current flow. If the circuit were purely resistive, there would be no phase difference between voltage and current. If the circuit were purely reactive there would be a 90° phase shift between voltage and current. The amount of phase shift is relative to the proportions of resistance and capacitive reactance present in the circuit.

7. The solution is found in phasor (addition) mathematics. For a solution to the voltage problem, draw a proportional phasor on the positive X-axis of

a plot that represents the voltage drop across the resistor and on the negative Y-axis draw a proportional phasor representing the voltage drop across the capacitor (this represents the 90° phase shift). Then draw a phasor representing the hypotenuse of the two right angles that you just plotted. This hypotenuse can be measured to get a relative answer or the Pythagorean method can be used to get a more exact answer. Use both methods to see if they agree. Using the phasor method, the calculated voltage V_A =

_____. Using the Pythagorean method, the calculated voltage V_A =

_____. Your answers should be approximately the same as indicated by the voltage source, V_A = 10 V.

8. As in inductive circuits, there are three methods to determine total opposition to current flow (Z) in an RC circuit: the Ohm's law method ($Z = V_A/I_T$), the phasor method, and the method using Pythagorean's theorem.

9. Now you will use all three methods. First, use the Ohm's law method. Impedance according to Ohm's law, Z = _____.

10. Now you need to determine the value of capacitive reactance (X_C). To calculate X_C use the formula:

$$X_C = \frac{1}{2\pi f C}$$

Calculated X_C = _____.

11. Now draw a proportional phasor on the X-axis of a plot representing the resistance of the resistor and then draw a proportional phasor representing the calculated capacitive reactance of the capacitor (this represents the 90° phase shift). Finally, draw a phasor representing the hypotenuse of the two right angles on the X-Y plot. Measure this hypotenuse to get a relative answer. Impedance according to the phasor method, Z = _____.

12. Finally, the Pythagorean method of determining impedance can be used to get a more exact answer. This answer should be close to the total opposition to current flow that was previously calculated by the other two methods. The Pythagorean formula is:

$$Z = \sqrt{R^2 + X_C^{\,2}}$$

Impedance according to the Pythagorean method, Z = _____.

13. Open circuit file **15-01b**. Activate the circuit and determine Z, X_C, and V_A. In this circuit, you need to calculate capacitive reactance first and then impedance. Calculated X_C = _____ and Z = _____.

14. Measure the voltage drops across the resistor and the capacitor with the DMM and then use the measurements to determine V_A using the Pythagorean method. Calculated (Pythagorean method) $V_A = $ _____.

15. Open circuit file **15-01c**. Activate the circuit and determine I_T, Z, and V_A. In this circuit, calculate capacitive reactance first and then impedance. Calculated $X_C = $ _____ and Z = _____.

16. Next, V_A can be determined by measuring the voltage drops across the two components and then using the Pythagorean method of solution. Measured $V_{R1} = $ _____ and $V_{C1} = $ _____. Calculated (Pythagorean method) $V_A = $ _____.

17. Finally, I_T can be determined by Ohm's law using the calculated circuit voltage (V_A) and impedance ($I_T = V_A/Z$). Calculated $I_T = $ _____.

18. Open circuit file **15-01d**. Activate the circuit and determine X_C, Z, I_T, and V_A. Calculated $X_C = $ _____, Z = _____, $I_T = $ _____, and $V_A = $ _____.

19. Open circuit file **15-01e**. Activate the circuit and determine circuit parameters to fill in Table 15-1. Verify the R_1 calculation with the DMM. The measured value of $R_1 = $ _____.

	Current	**Voltage**	**Opposition to Current Flow**
R_1			R =
C_1			$X_C = $
Totals			Z =

Table 15-1   Parameters of a Series RC Circuit

20. Open circuit file **15-01f**. Activate the circuit and determine circuit parameters to fill in Table 15-2 on page 146. Verify the R_1 calculation with the DMM. Measured $R_1 = $ _____.

	Current	Voltage	Opposition to Current Flow
R_1			$R =$
C_1			$X_C =$
Totals			$Z =$

Table 15-2   Parameters of a Series RC Circuit

● *Troubleshooting Problems:*

21. Open circuit file **15-01g**. The problem with this circuit is that the current is extremely low. Determine the values of X_C and Z first and then I_T. Calculated

    $X_C =$ _____, $Z =$ _____, and $I_T =$ _____.

22. Activate the circuit and use the DMM to determine the circuit fault (assume that the capacitor is the correct value according to the schematic). The

    problem is _____

    _____.

23. Open circuit file **15-01h**. There is a problem with this circuit; the resistor is smoking, and too much current is being drawn from the power source. Determine the values of X_C and Z and then calculate the "should be" value of

    I_T. Calculated $X_C =$ _____, $Z =$ _____, and $I_T =$ _____.

    Activate the circuit and use the DMM as an ohmmeter to measure the resistor

    and the capacitor. Measured value of $R_1 =$ _____ and $C_1 =$ _____. The

    problem is _____

    _____.

Activity 15.2: Parallel Resistive-Capacitive (RC) Circuits

1. As in parallel DC circuits, parallel RC circuits have identical voltage drops across every parallel component in the circuit (see Figure 15-2). The current will divide between the parallel branches according to the amount of opposition (resistance or capacitive reactance) that each parallel branch offers to the flow of current. Phase relationships are another matter. Differing from series RC circuits, parallel RC circuits have the same voltage across all of the parallel components. There is always a current and voltage phase shift across capacitors, so the current through capacitors must be shifted in phase relationships from the voltage by 90°.

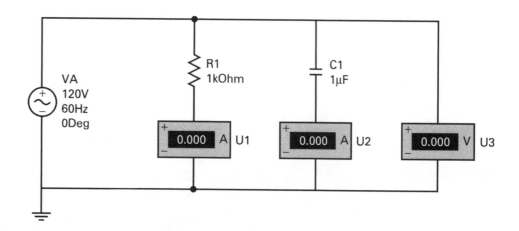

Figure 15-2   A Parallel RC Circuit

2. Open circuit file **15-02a**. This circuit displays the voltage drop across the parallel components and the current flow through the two parallel branches. Determine the sum of the current flow through the parallel branches. Activate the circuit and determine the arithmetic sum of the currents of the parallel branches ($I_{R1} + I_{C1}$). The arithmetic sum of the current in the parallel branches = _____.

3. Is the total of the branch currents equal to I_T as reflected by U_1? Obviously not. Measured I_T = _____.

4. Use phasor addition to calculate I_T with resistor current reflected on the positive portion of the X-axis and capacitor current reflected on the negative portion of the Y-axis. Circuit current using the phasor method, I_T = _____.

5. Then use the Pythagorean method to determine circuit current. The calculated current should equal the current displayed on meter U_1. Circuit current using the Pythagorean method, I_T = _____.

6. Now determine the reactance offered by the capacitive branch of the circuit. First, calculate X_C. Calculated X_C = _____.

7. Determine the impedance of the circuit by the Ohm's law method ($Z = V_A/I_T$). Calculated Z = _____.

8. In a parallel RC circuit, the Ohm's law method is the easiest method to determine impedance when current and voltage parameters are available, but there

is a mathematical method using the parallel resistance formulas and the Pythagorean theorem (see Figure 15-3). The formula to use is:

$$Z = \frac{R \times X_C}{\sqrt{R^2 + X_C^2}}$$

9. What is Z using this formula? Calculated Z = _____.

10. Open circuit file **15-02b**. Activate the circuit to determine circuit parameters and fill in Table 15-3.

	Current	Voltage	Opposition to Current Flow
R_1			$R_1 =$
R_2			$R_2 =$
RT			$R_T =$
C_1			$X_{C1} =$
C_2			$X_{C2} =$
C_T			$X_{CT} =$
Totals			$Z =$

Table 15-3   Parallel RC Circuit Parameters

● **Troubleshooting Problems:**

11. Open circuit file **15-02c**. Activate the circuit to determine circuit parameters and fill in Table 15-4. The resistors have the same value of resistance and should have the same amount of current flowing through them. Also, the two capacitors have the same value of capacitance and, thus, the same value of capacitive reactance, and should also have the same amount of current flowing through them. There appear to be two problems in this circuit. Find them.

12. The problems are _____

_____

_____

_____.

	Current Should Be	Current Is	Voltage Is	Opposition to Current Flow
R_1				$R_1 =$
R_2				$R_2 =$
RT				$R_T =$
C_1				$X_{C1} =$
C_2				$X_{C2} =$
C_T				$X_{CT} =$
Totals				$Z =$

Table 15-4 Parallel RC Circuit Parameters

Activity 15.3: Series-Parallel Resistive-Capacitive (RC) Circuits

1. A series-parallel RC circuit is a combination of series and parallel RC sections where series and parallel circuit parameters operate (see Figure 15-3). Each parallel RC section has to be simplified until it is minimized, hopefully,

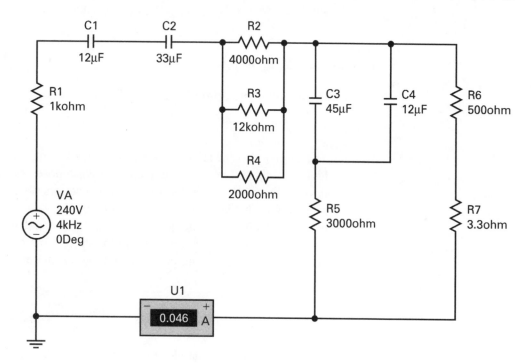

Figure 15-3 A Series-Parallel RC Circuit

into simple series R and C sections. Then, further simplification can take place with as many series components as possible.

2. Open circuit file **15-03a**. Activate the circuit and record the circuit current.

   Measured $I_T =$ _____.

3. Calculate the necessary combinations to simplify the circuit. Combine resistors R_6 and R_7 into combination resistor R_{6-7} and resistors R_1, R_2, R_3, and R_4 into combination resistor $R_{1-2-3-4}$. Combine capacitors C_1 and C_2 into combination capacitor C_{1-2} and capacitors C_3 and C_4 into combination capacitor C_{3-4}.

   $R_{1-2-3-4} =$ _____, $R_{6-7} =$ _____, $C_{1-2} =$ _____, and

   $C_{3-4} =$ _____.

4. Open circuit file **15-03b**. Change the value of the resistors and capacitors to reflect the various combinations of components in the previous circuit (15-03a). Activate the circuit and record the circuit current. Measured $I_T =$

   _____.

5. The current should be the same as in the previous circuit if your calculations were correct and correctly placed in the circuit.

6. Open circuit file **15-03c**. Simplify this circuit into a single resistor and a single capacitor. $R_{SIMPLIFIED} =$ _____ and $C_{SIMPLIFIED} =$ _____.

● *Troubleshooting Problems:*

7. Open circuit file **15-03d**. Activate the circuit and record I_T. Measured $I_T =$

   _____.

8. The circuit current is supposed to be 360–370 mA and it is considerably higher. In this circuit, C_1 is defective; it is supposed to be 10 μF. In looking through the parts bins in the shop, four capacitors were located that look the same as C_1. This is a very critical circuit and the value of the capacitor has to be exact (within tolerance). There is no 10μF-capacitor in the assortment. Take another look at the circuit and notice that there is a way to place the proper capacitance in the circuit using one capacitor. What would you do?

   I would _____

   _____ using capacitor(s) _____. Go ahead; modify the circuit.

9. Activate the circuit; what does I_T measure now? Measured $I_T =$ _____.

10. After replacing the part and turning on the equipment, it runs for a few hours and then a "poof" occurs; the end shoots out of the capacitor(s) and tin foil shoots across the room. The replacement part is history. In continuing the troubleshooting process, another bad component in the equipment is found that caused the replacement capacitor(s) to go bad and you replace it. Now it is necessary to temporarily fix the circuit with substitute components to get the equipment operational while the correct parts are on order. Is there any series-parallel combination using the other capacitors that will make it possible to temporarily replace C_1 (and C_2, if necessary) until the correct replacement part arrives?

11. After checking out the available parts situation, the best solution is to use _____. Now, change the parts according to your solution and check the circuit. Measured $I_T =$ _____.

12. Open circuit file **15-03e**. Calculate X_C and the proper value of I_T. Then activate the circuit and record X_C and I_T. Calculated $X_C =$ _____ and $I_T =$ _____. Measured $I_T =$ _____.

13. Obviously, there is a problem. Find the faulty component (a capacitor) and use the assortment of spare capacitors to replace it. Explain how you would connect the replacement capacitor(s). I would _____.

14. Activate the circuit and determine if it is operating properly. I_T should be the same as your calculation for I_T in Step 12 (assuming that your calculation was correct). Measured $I_T =$ _____.

Activity 15.4: Pulse Response and Time Constants of RC Circuits

1. Up to this point our study of RC circuits has been under sinusoidal signal conditions. Next, we will study the operation of RC circuits under non-sinusoidal conditions, primarily square waves, but also the unique situation in DC circuits at turn-on and turn-off time. The operation of an RC circuit under squarewave conditions is similar to the turn-on and turn-off situation in a DC circuit where there is a sudden transition from off-to-on or on-to-off. RC circuits do react to these sudden changes that take place in a DC circuit, but after the initial reaction, the capacitor performs the function of blocking DC current flow only.

2. Open circuit file **15-04a**. Activate the circuit and view the increase in V_A from 0 V to 10 V over five time constants. Calculate the duration of 1 TC. At the end of 5 TC, what is the voltage across the capacitor? Calculated 1 TC = _____. Voltage across C_1 after 5 TC = _____.

3. After viewing rise time, toggle the Switch J1 down (space bar) to view fall time. Fall time should be a reverse image of rise time. Notice that R_2 is set for zero kOhms in value. Change the value to 20 kOhms and observe the change in fall time. What did you observe and why? What is 1 TC for fall time? _____

    _____.

    1 TC for fall time = _____.

4. Determine the voltage at the end of 40 milliseconds (four major divisions) from the start of the rise time. The voltage after 40 msec = _____.

5. How many time constants is 30 msec? What is the voltage across C_1 after 30 msec? 30 msec = _____ TC. Measured voltage across C_1 after 30 msec =

    _____.

6. Place the red or blue cursor on the 20-msec point and read the voltage directly from the related window. The voltage as indicated by the "red cursor" window at the 20-msec point is _____.

7. Once again, the oscilloscope has proven to be a powerful tool for viewing the inner workings of an electronic circuit. When displaying the V_R and V_C wave-shapes simultaneously, the two curves would cross each other in a manner similar to the Universal time-constant chart.

8. Open circuit file **15-04b**. Determine the voltage rise and fall times in time constant intervals and enter the data in Table 15-5.

Time Constant	1 TC	2 TC	3 TC	4 TC	5 TC
Rise Time Voltage Reading					
Fall Time Voltage Reading					

Table 15-5 Voltage Rise and Fall Time in TC Intervals

9. RC circuits have a predicable rise and fall time when sudden transitions occur in a DC circuit. The next subject under study is the response of an RC circuit to the fast rise and fall time of square waves (this is called "step response"). There are two basic types of capacitor action in response to square waves; they are integration and differentiation. The method of placement of capacitors in a circuit determines this response. Figure 15-4 displays the two methods of placement.

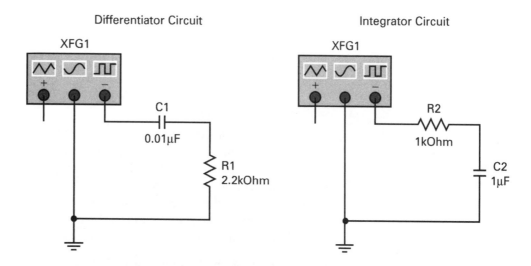

Figure 15-4   Capacitor Placement for Integration and Differentiation

10. Open circuit file **15-04c**. Activate the circuit and observe the step response of an RC circuit to a square wave. The response across the capacitor is being compared to the input square wave. The upper red trace is the square wave input and the lower blue trace is the "integrated" response of the capacitor to the signal. Calculate the duration of five time constants in this circuit?

    Calculated 5 TC = _____msec.

11. Does the Channel B signal ever reach the 5-TC point in the display? Yes_____ or No_____. If you changed the value of R_1 or C_1 to shorten the time constant would the integrated response to the square wave reach the 5-TC point? Yes_____ or No_____.

12. Change R_1 to 5 kΩ and observe the waveshape. Does the Channel B signal reach the 5-TC point on the display? Yes_____ or No_____. Return R_1 to 30 kΩ and change C_1 to 2 µF. Observe the waveshape. Does the Channel B signal reach the 5-TC point on the display? Yes_____ or No_____.

13. Open circuit file **15-04d**. Activate this integrator circuit and observe the step response in two modes of integrator action. The display for the long time constant mode is the type of signal that is usually considered to be an output of an integrator circuit. The display for the short time constant circuit has some integrator action, as can be seen from the rounding off of the leading and lagging edges of the square wave. The only difference between the two time constant circuits is in the size of the resistor. If the resistance value of R_2 were

made larger, then the integration action would be less observable. Calculate the duration of one time constant (1 TC) for each of the two time constant options in this circuit. For the Short Time Constant option: 1 TC = _____.

For the Long Time Constant option: 1 TC = _____.

14. Open circuit file **15-04e**. Activate this differentiator circuit and observe the step response of a capacitor in series with the square wave input to the circuit. As with the integrator circuit, the response (sharpness) of the differentiated square wave is dependent on the values of the capacitor and the load resistor. If the resistance value of R_1 were made smaller, then the differentiation action would be sharper (the differentiated signal would be of shorter duration). Also, if the capacitance value of the capacitor were made smaller, the differentiated signal duration would be less. What is the duration of the one time constant in this circuit? Calculated 1 TC = _____ μsec.

● *Troubleshooting Problem:*

15. Open circuit file **15-04f**. Activate the circuit. In assembling this circuit for the engineering department at work, the requirement for an extremely short time constant resulting in a differentiated pulse that looks like (the way it is set up on the oscilloscope) a spike became a necessity. It is now necessary to choose between two options: (1) to increase/decrease (circle the best answer) the resistance of the resistor, or (2) to increase/decrease (circle the best answer) the capacitance of the capacitor. Which would be the least expensive?

Probably to _____ would be least expensive.

16. Resistive-Inductive-Capacitive Circuits

References

Electronics Workbench®, *MultiSIM* Version 7

Electronics Workbench®, *MultiSIM* Version 7 User's Guide

Objectives After completing this chapter the student should be able to:

- Operate series circuits containing resistors, inductors, and capacitors
- Operate parallel circuits containing resistors, inductors, and capacitors
- Operate series-parallel circuits containing resistors, inductors, and capacitors
- Determine resonant frequencies of RLC circuits
- Troubleshoot resistive-inductive-capacitive circuits

Introduction

Resistive-inductive-capacitive (RLC) circuits combine resistors, inductors, and capacitors in various circuit configurations. The activities that take place in a circuit of this type are a combination of all three characteristics: inductance, capacitance, and resistance. The determination of whether the circuit is primarily resistive, inductive, or capacitive depends on the relationships between the three types of components. If one of the three characteristics offers more or less opposition to current action than the other two, the characteristics of the circuit are modified accordingly and the circuit tends to be resistive, inductive, or capacitive depending on which of the three is dominant.

When analyzing RLC circuits, we discover that these circuits are various combinations of resistors, inductors, and capacitors with aiding and opposing interactions in the circuit. There is no voltage and current phase shift with the resistive elements, yet there are 90° voltage and current phase shifts with the capacitors and inductors. And, those phase shifts are in opposition to one another. It is important to remember that the current flow is the same everywhere in a series circuit: through the resistor, the inductor, and the capacitor. While the voltage and current through the resistor is in phase, the current lags the voltage by 90° through the inductor, and the current leads the voltage by 90° through the capacitor.

The parallel and series-parallel circuits offer even more complexity. To solve for circuit parameters with these more complex circuits, use the same methods presented in earlier chapters, the methods of phasor addition along with the Pythagorean theorem, Ohm's law, and Kirchhoff's voltage and current laws.

Activity 16.1: Series Resistive-Inductive-Capacitive (RLC) Circuits

1. The series **resistive-inductive-capacitive (RLC)** circuit contains resistors, inductors, and capacitors connected in series as loads. In Figure 16-1, a series RLC circuit is displayed. This series circuit contains an AC signal source, a resistor, an inductor, and a capacitor.

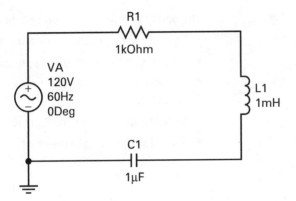

Figure 16-1   A Series RLC Circuit

2. Open circuit file **16-01a**. In this series RLC circuit, ammeter U_1 is installed to measure series current. Remember, according to the definition of a series circuit, current flow is the same everywhere in a series circuit. Also, the resistive and reactive components drop voltages in proportion to their resistance or reactance. Activate the circuit and determine current flow. Measured

   $I_T =$ _____.

3. Measure the voltage drops across the resistor, the inductor, and the capacitor.

   Measured $V_{R1} =$ _____, $V_{L1} =$ _____, and $V_{C1} =$ _____.

4. Does the sum of the voltage drops equal the voltage applied to the circuit by the voltage source? The sum of the voltage drops (<u>does/does not</u>) equal the voltage applied to the circuit. As has been determined in the study of reactive circuits, there are definite interactions and differences between resistive and reactive circuits.

5. In this type of circuit the phase shifts between the voltage and current of the reactive components are to be expected. The first factor to consider at this point is the differences in the phase shift interactions between the inductors and the capacitors. On one hand, the reactive phase shift has current leading voltage (capacitive), and on the other, current lags the voltage (inductive). Ultimately, the amount of phase shift in either direction will depend upon the degree of opposition offered to the circuit by the inductive reactance and the capacitive reactance. The reactance that is dominant, inductive or capacitive, will establish the characteristic of the circuit. Then, in conjunction with resistance, the circuit will act as an R_L or an R_C circuit. The formula for calculating impedance in a circuit of this type is:

$$Z = \sqrt{R^2 + (X_L - X_C)^2}$$

6. What is the impedance of this circuit using this formula? First, calculate X_L and X_C and then calculate impedance. Calculated $X_L =$ _____ and $X_C =$ _____. Calculated (impedance formula method) $Z =$ _____.

7. The Ohm's law formulas are still the easiest methods to use to determine circuit factors such as impedance, current, or voltage. Determine the amount of opposition (impedance) to current flow in this circuit according to Ohm's law $(Z = V_A/I_T)$. The impedance $(Z) =$ _____.

8. The solution set to the reactance and impedance in this circuit is found in phasor mathematics and the Pythagorean theorem. To solve for circuit impedance, start by drawing a proportional phasor on the X-axis of a plot, representing the resistance of the resistor. On the negative portion of the Y-axis, draw a proportional phasor representing the capacitive reactance of the capacitor (this represents one of the 90° phase shifts). On the positive portion of the Y-axis draw a proportional phasor representing the inductive reactance of the inductor (this represents the other 90° phase shift). Next, algebraically sum up the phasors on the Y-axis, subtracting X_C from X_L with dominance being established by the higher value (absolute value) of reactance. Whichever absolute value is larger indicates whether the circuit is ultimately R_C or R_L. Finally, draw a phasor representing the hypotenuse of the two right angles on the plot (resistance and reactance). This hypotenuse can be measured to get a relative answer by the phasor method, or the Pythagorean method can be used to get a more exact answer. What is the impedance of the circuit according to both methods? Calculated (phasor method) $Z =$ _____. Calculated (Pythagorean method) $Z =$ _____.

9. Open circuit file **16-01b**. Determine circuit impedance Z, X_C, X_L, and V_A. In this circuit the first step is to measure the voltage drops. Measured $V_R =$ _____, $V_L =$ _____, and $V_C =$ _____.

10. Use these voltage drops to determine V_A (phasor or Pythagorean method) and then with the circuit current measurement to calculate impedance. Calculate inductive reactance (X_L) and capacitive reactance (X_C). Calculated $V_A =$ _____ and $Z =$ _____. Calculated $X_L =$ _____ and $X_C =$ _____.

11. Open circuit file **16-01c**. Activate the circuit and determine I_T, Z, and V_A. In this circuit, there are several directions that can be taken. Your first choice might be to solve for inductive and capacitive reactance and impedance and then measure the voltage drops to determine applied voltage with the Pythagorean method, to calculate circuit current by measuring the voltage drop across the resistor and then, calculating I_T, follow up by voltage drop measurements across the reactive components, and finally, calculate reactance using the known current flow. Choose the best method. Calculated $X_L =$ _____, $X_C =$ _____, and $Z =$ _____.

12. Next, calculate V_A by measuring the voltage drops across the components, using the Pythagorean method of solution. Measured $V_R =$ _____, $V_L =$ _____, and $V_C =$ _____. Calculated (Pythagorean method) $V_A =$ _____.

13. Finally, calculate I_T using Ohm's law. Calculated $I_T =$ _____.

14. Open circuit file **16-01d**. Activate the circuit and enter parameters in Table 16-1. Measure V_R, V_L, and V_C. Measured $V_R =$ _____, $V_L =$ _____, and $V_C =$ _____.

	Current	Voltage	Opposition to Current Flow
R_1			$R =$
L_1			$X_L =$
C_1			$X_C =$
Totals	$I_T =$	$V_A =$	$Z =$

Table 16-1   RLC Series Circuit Data

15. Open circuit file **16-01e**. Determine circuit parameters and enter your data into Table 16-2. Determine the value of R_1 with the DMM. The measured value of $R_1 =$ _____.

	Current	**Voltage**	**Opposition to Current Flow**
R_1			R =
L_1			X_L =
C_1			X_C =
Totals	I_T =	V_A =	Z =

Table 16-2   Parameters of a Series RLC Circuit

16. Open circuit file **16-01f**. Determine the circuit parameters and fill in Table 16-3. Determine the value of R_1 with the DMM. Remember that meter U_1 will slightly affect the resistance measurement. To get a more accurate measurement of the resistor, temporarily disconnect the meter. The measured value of R_1 = _____.

	Current	**Voltage**	**Opposition to Current Flow**
R_1			R =
L_1			X_L =
C_1			X_C =
Totals	I_T =	V_A =	Z =

Table 16-3   Parameters of a Series RLC Circuit

● *Troubleshooting Problems:*

17. Open circuit file **16-01g**. There is a problem with this circuit; the current reading is indecisive. Calculate the values of X_L, X_C , and Z and then the "should be" value of I_T. What is the problem with the ammeter reading? Calculated X_L = _____, X_C = _____, Z = _____, and I_T = _____. The problem with the circuit is _____.

18. Fix the problem, reactivate the circuit, and measure I_T again. Measured I_T = _____. The reading should match your calculation.

19. Open circuit file **16-01h**. There is a problem with this circuit. The current is less than half of its proper value. Determine X_L, X_C , and Z and then

calculate the proper value of I_T. Calculated $X_L =$ _____, $X_C =$ _____, and $Z =$ _____. The proper value of $I_T =$ _____.

20. Use the test equipment to determine the circuit fault. *Hint:* Measure the voltage drops. $V_R =$ _____, $V_L =$ _____, and $V_C =$ _____. The problem is _____

_____

_____.

Activity 16.2: Parallel Resistive-Inductive-Capacitive (RLC) Circuits

1. Parallel RLC circuits have identical voltage drops across every parallel component in the circuit (see Figure 16-2) just like the resistive parallel circuits that we have studied. The current will divide between the parallel branches according to the amount of opposition (resistance or reactance) that each parallel branch offers to the flow of current. Phase relationships are another matter. Reactance in a circuit always causes phase shift of current and voltage. As a result, the current through the reactances is phase shifted from the voltage by 90°. This is similar to the RLC series circuits; the reactance that dominates in conjunction with the resistance sets the characteristics of the circuit.

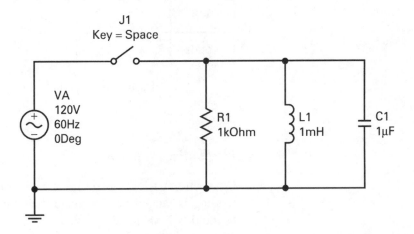

Figure 16-2   A Parallel RLC circuit

2. Open circuit file **16-02a**. This circuit demonstrates the voltage drops across the parallel components and the current flow through each component. Add the individual branch currents.

The sum of branch currents = _____. Is this sum reflected by ammeter U_1? Obviously not.

3. Now use the phasor and Pythagorean methods to calculate I_T. Resistor current is reflected on the positive portion of the X-axis, inductor current on the positive portion of the Y-axis, and capacitor current is reflected on the negative portion of the Y-axis. Calculate I_T and then verify that the calculated I_T is equal to the current displayed on meter U_1. Calculated I_T = _____. Measured I_T = _____.

4. Determine the inductive and capacitive reactance offered by the respective branches of the circuit. Calculated X_L = _____ and X_C = _____.

5. Determine the impedance of the circuit by the Ohm's law method. Measured Z = _____.

6. Open circuit file **16-02b**. Determine circuit parameters and fill in Table 16-4.

	Current	**Voltage**	**Opposition to Current Flow**
R_1			$R_1 =$
R_2			$R_2 =$
R_T			$R_T =$
L_1			$X_{L1} =$
L_2			$X_{L2} =$
L_T			$X_{LT} =$
C_1			$X_{C1} =$
C_2			$X_{C2} =$
C_T			$X_{CT} =$
Totals			$Z =$

Table 16-4   Parallel RLC Circuit Parameters

● *Troubleshooting Problem:*

7. Open circuit file **16-02c**. Determine circuit parameters and fill in Table 16-5. The resistors have the same value of resistance and should have the same amount of current flowing through them. The two capacitors are the same and should have the same amount of current flowing through them. Finally, the

two inductors are also the same and should have the same current flowing through them. There appear to be two problems in this circuit. Find them. The

problems are _____

_____ and _____

_____.

	Current Should Be	Current Is	Voltage Is	Opposition to Current Flow
R_1				$R_1 =$
R_2				$R_2 =$
R_T				$R_T =$
L_1				$X_{L1} =$
L_2				$X_{L2} =$
L_T				$X_{LT} =$
C_1				$X_{C1} =$
C_2				$X_{C2} =$
C_T				$X_{CT} =$
Totals				$Z =$

Table 16-5   Parallel RLC Circuit Parameters

Activity 16.3: Series-Parallel Resistive-Inductive-Capacitive (RLC) Circuits

1. As in resistive series-parallel circuits, series-parallel RLC circuits are a combination of series sections where series-circuit parameters operate and parallel sections where parallel-circuit parameters operate (see Figure 16-3). The first objective in analyzing these types of circuits is to simplify the circuits as much as possible by combining individual components into single entities, joining series components, calculating parallel combinations, and finally reaching a point where no further simplification is possible.

2. Open circuit file **16-03a**. Simplify this circuit by calculating the necessary combinations. Combine resistors R_4 and R_5 into combination resistor R_{4-5}

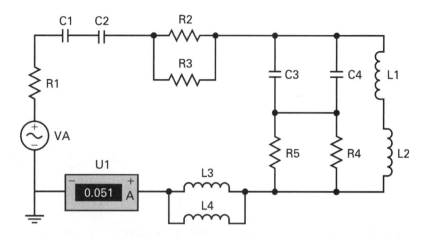

Figure 16-3   A Series-Parallel RLC Circuit

and resistors R_1, R_2, and R_3 into combination resistor R_{1-2-3}. Combine capacitors C_1 and C_2 into combination capacitor C_{1-2} and capacitors C_3 and C_4 into combination capacitor C_{3-4}. Finally, combine inductors L_1 and L_2 into combination inductor L_{1-2} and inductors L_3 and L_4 into combination inductor L_{3-4}. The values of $R_{4-5} = $ _____, $R_{1-2-3} = $ _____, $C_{1-2} = $ _____, $C_{3-4} = $ _____, $L_{1-2} = $ _____, and $L_{3-4} = $ _____.

3. Activate the circuit and record the circuit current. Measured $I_T = $ _____.

4. Open circuit file **16-03b**. Change the value of the resistors and capacitors to reflect the simplification of the series and parallel sections of the previous circuit. Activate the circuit and record the circuit current. The ammeter should display the same value for I_T as the previous circuit if the combinations were correctly calculated. Measured (after simplification) $I_T = $

    _____.

5. Open circuit file **16-03c**. Simplify this circuit into a single resistor, a single inductor, and a single capacitor. $R_{SIMPLIFIED} = $ _____, $L_{SIMPLIFIED} = $ _____, and $C_{SIMPLIFIED} = $ _____.

6. Calculate X_L, X_C , Z, and I_T. Calculated $X_L = $ _____, $X_C = $ _____, $Z = $ _____, and $I_T = $ _____.

● ***Troubleshooting Problems:***

7. Open circuit file **16-03d**. This circuit has three problems in it: one shorted resistor, one leaky inductor, and one open capacitor. Activate the circuit and

record I_T (I_T should be 1.547 mA). Then locate the defective parts and replace them with spare parts located to the right of the circuit. Measured I_T = _____. The defective components are R_____, L_____, and C_____.

8. Change the values of the replacement components to the rated values of the defective components that are being replaced. Activate the circuit and record the new value for I_T. Measured (after repair) I_T = _____.

Activity 16.4: Series and Parallel Resonant Circuits

1. There is a point of operation in a series RLC circuit called **resonance**. This point of operation occurs when the impedance of the circuit consists of resistance only; at this point inductive reactance is equal to capacitive reactance ($X_L = X_C$). At resonance, the inductive reactance and the capacitive reactance cancel each other out on an X-Y plot. At resonance, frequency is the determining factor; the inductive reactance is equal to the capacitive reactance at one frequency only. The voltage drops across the inductor and the capacitor are equal, and the current is at a maximum point. The frequency at which this condition occurs is called the **resonant frequency** (f_r). In every series AC circuit containing inductors and capacitors, there is a resonant frequency. You need to remember that, at resonance, the voltage drop across the inductor is equal to the voltage drop across the capacitor and they are at their maximum point. The formula for resonant frequency is:

$$f_r = \frac{0.159}{\sqrt{LC}}$$

2. Open circuit file **16-04a**. In this series RLC circuit, the formula for the frequency of resonance tells us that the resonant frequency is 5035.5 Hz (see Figure 16-4). Set the frequency of the signal source to the specified frequency. As previously stated, at the frequency of resonance the impedance of the circuit is equal to the resistance of the circuit alone. In this circuit, the value of the resistor is 1 kΩ. Calculate circuit current at resonance. Calculated I_T at $f_r = V_A/R_1$ = _____.

3. Activate the circuit and verify the current calculation. Measured I_T at f_r = _____.

4. Open circuit file **16-04b**. What is the resonant frequency of this circuit and what is circuit current at that point? Circuit current at f_r is always $I_T = V_A/R_1$ (total resistance of the circuit). Calculated f_r = _____ and I_T = _____.

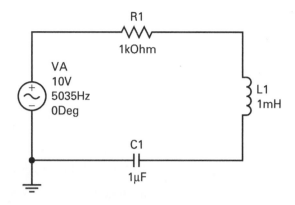

Figure 16-4   A Series RLC Circuit at the
Resonant Frequency

5. Adjust the frequency of the signal source to f_r and activate the circuit. The actual circuit current flow should be close to the calculated value. Measured

   $f_r =$ _____ and $I_T =$ _____.

6. In many aspects, and somewhat to be expected, the parameters for a parallel RLC circuit operate in a manner opposite a series RLC circuit. Circuit impedance is at its maximum value rather than at its minimum. At the resonant frequency the circuit is resistive because the capacitive and inductive reactances have cancelled each other out; the current flow in the reactive branches is equal and opposite. Circuit current is at a minimum value (see Figure 16-5) and is equal to the current in the resistive branch.

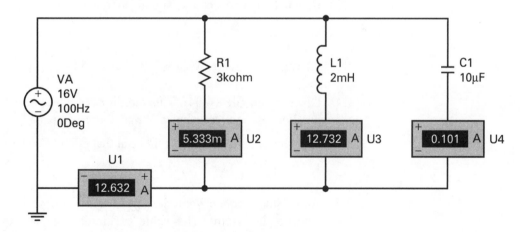

Figure 16-5   A Parallel RLC Circuit at the Resonant Frequency

7. Open circuit file **16-04c**. In this parallel RLC circuit, the formula for the frequency of resonance (usually the same formula that is used with series circuits) tells us that the resonant frequency is 1124.3 Hz. Set the frequency of the signal source to that frequency. As previously stated, at the frequency

of resonance the impedance of this parallel circuit is only equal to the resistance of the circuit and is at its maximum point. In this circuit, the value of resistor is 100 Ω. Calculate circuit current ($I_T = V_A/R_1$) at resonance. Calculated I_T at $f_r = V_A/R_1 = $ _____.

8. Activate the circuit and verify the current calculation. Measured I_T at $f_r = $

    _____.

9. Open circuit file **16-04d**. Determine the resonant frequency of this circuit and calculate I_T at f_r. Calculated $f_r = $ _____ and I_T (at f_r) = _____.

10. Adjust the frequency of the signal source to f_r and activate the circuit. The actual circuit current flow should be close to the calculated value. Measured I_T (at f_r) = _____.

● **Troubleshooting Problems:**

11. Open circuit file **16-04e**. This circuit does not appear to be operating at series resonance; current at f_r should be 200 mA. What is the problem with the circuit? The problem is _____

    _____

    _____.

12. Calculate f_r and set the signal source to this frequency. Calculated $f_r = $

    _____.

13. Activate the circuit. What is I_T? Measured $I_T = $ _____.

14. Open circuit file **16-04f**. This circuit does not appear to be operating at the parallel resonant frequency. What is the resonant frequency of this circuit and what should circuit current be at that point? Calculated $f_r = $ _____ and $I_T = $ _____.

15. Adjust the frequency of the signal source to your calculation for f_r and activate the circuit. The value of circuit current flow should match the calculation. Measured I_T (at f_r) = _____.

17. Transformer Circuits

References

Electronics Workbench®, MultiSIM Version 7

Electronics Workbench®, MultiSIM Version 7 User's Guide

Objectives   After completing this chapter the student should be able to:

- Operate and analyze circuits containing transformers
- Use step-up and step-down transformers
- Determine turns ratios for impedance matching
- Troubleshoot transformer circuits

Introduction

Transformers are standard components in electronics and are usually very small for printed circuit board mounting and similar electronics applications. At the same time, transformers can weigh tons and be so large that cranes and heavy-duty construction equipment are required to install and service them. In the first type of application, the transformer would handle low voltages and currents (or at least low currents). In the larger application, the transformer may be used for voltage distribution, controlling voltages in the hundreds of kilovolts and currents in the thousands of amperes. In any location where there is electrical power and electronic circuits, there are usually transformers. They are one of the most versatile components available for changing AC voltages and currents. Technicians and maintenance personnel must understand transformers and transformer circuits of all types; they have to be able to troubleshoot and analyze circuits and equipment containing transformers.

Activity 17.1: Transformer Circuits

1. **Transformers** operate on the principle of mutual inductance, a magnetic linkage that uses electromagnetic principles to transfer energy from one coil circuit to an adjacent coil circuit. In MultiSIM, the transformers are very basic, performing necessary functions to provide transformer action in circuits. Figure 17-1 on page 168 displays some of the types of transformers used in MultiSIM.

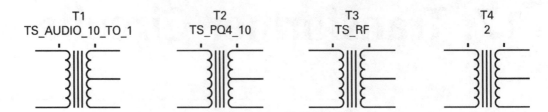

Figure 17-1 Transformers Used in MultiSIM

2. The **basic (virtual) transformer** is just that—basic. It provides several of the necessary functions of a transformer, voltage step-up and voltage step-down. The isolation function of transformers is not available in MultiSIM. To properly simulate, both sides of the transformer require a **common reference point** (usually ground), which takes away the isolation function. *Note:* You must ground both sides of a transformer or provide a common reference point.

3. When discussing the step-up and step-down properties of transformers, we specifically refer to the step-up and step-down of voltages and not current. In a transformer circuit, current changes in a proportional ratio that is the opposite of the voltage ratio. For example, across the transformer, from primary to secondary, if voltage steps up, then current steps down, and if voltage steps down, then current steps up. Excluding core losses in a transformer, power of the primary is equal to the power of the secondary ($P_{prim} = P_{sec}$).

4. Open circuit file **17-01a**. The default for the virtual transformer is a voltage step-down at a 2:1 ratio; the voltage of the secondary is one half that of the primary. Activate the circuit and observe the step-down from primary to secondary. Also observe that, on the secondary side of this transformer, there is a center tap that is a point of connection where the secondary voltage is halved. This is a very necessary feature of transformers that are used in power supply circuits in electronics equipment. Verify the step-down function of the primary and secondary circuits of the transformer with voltmeters U_1 (primary circuit) and U_2 (secondary circuit). Measured

   Primary Voltage = _____. Measured Secondary Voltage = _____.

5. Notice that the transformer has the number 2 below the reference ID (T_1). This number indicates that the voltage in the primary is two times higher than the voltage in the secondary, in other words, a 2:1 ratio. To change the transformer step-down or step-up ratio, access the component menu and choose the **Value** tab (see Figure 17-2).

6. To change the transformer ratio, change the number in the **Primary-to-Secondary Turns Ratio (N)** box to the desired number of primary turns

Figure 17-2 Using the Value Tab to Change Turns Ratios

in relationship to the number 1 for the secondary. For example, placing a 5 in that box creates a 5:1 turns ratio (step-down); the number 10 would create a 10:1 turns ratio (step-down). If you want to step up a voltage, then place a number in the window that is less than one that will proportionally provide a stepped-up voltage in the secondary. For example, 0.5 in the window would provide a 1:2 step-up ratio and the number 0.1 in the window would provide a 1:10 step-up ratio. A formula to help with step-down ratios is:

$$N = \frac{V_{Primary}}{V_{Secondary}}$$

7. Now edit the basic transformer in this circuit so that the secondary voltage is 480 V (the input is 120 V) for a step-up of 1:4. What entry would have to be entered in the Primary-to-Secondary Turns Ratio (N) edit block in the transformer menu to accomplish the task? Make the entry in the N block and measure the secondary voltage. The Primary-to-Secondary Turns Ratio block would be set to _____. Measured secondary voltage = _____.

8. Open circuit file **17-01b**. Use the DMM to determine the turns ratio of this transformer. The turns ratio is _____. The Primary-to-Secondary Turns Ratio block is set to _____.

9. Open circuit file **17-01c**. Measure the primary voltage first; then edit the turns ratio of the transformer so that the secondary voltage is 50 V. Then measure the secondary voltage. Measured Primary Voltage = _____. The new Primary-to-Secondary Turns Ratio setting is _____. Measured Secondary Voltage = _____.

10. Open circuit file **17-01d**. Measure the primary voltage first; then edit the transformer so that the output voltage is 200 V. Measure the secondary voltage to verify the setting. Measured Primary Voltage = _____. The new Primary-to-Secondary Turns Ratio setting is _____. Measured Secondary Voltage = _____.

11. Open circuit file **17-01e**. Edit the transformer so that the output voltage of the secondary is 8 V. The entry in the **Primary-to-Secondary Turns Ratio** (N) block of the transformer edit menu would be edited to _____ which would be a ratio of _____. Measured Secondary Voltage = _____.

12. Open circuit file **17-01f**. In this circuit the secondary center tap is going to be checked out. The transformer is set for a 2:1 step-down. The secondary voltage from top to bottom of the secondary should be one-half of the primary voltage. The voltage from the center tap (TPA) to either end of the secondary winding should be one-half of the secondary voltage. Measured Secondary Voltage = _____. Center tap (TPA) to the top of the secondary = _____. Center tap (TPA) to the bottom of the secondary = _____.

13. As mentioned previously, whatever the voltage does in the secondary is reflected in an opposite manner by the current. If the voltage of the secondary steps-up by a factor of two, then the current capabilities decrease and are cut in half. If the secondary voltage steps down to one-half of the primary voltage, then the secondary current capabilities steps up by two. This is modified by a number of factors:

    a. There is always trickle current in the primary that is present for low levels of current usage by a load.

    b. There is always a power loss (primarily core losses) from primary to secondary (usually less than 5%).

    c. The current drain of the secondary is always reflected back to the primary current drain in the appropriate step-up or step-down ratio.

    d. The power of the primary is equal to the power of the primary excluding core losses.

● *Troubleshooting Problem:*

14. Open circuit file **17-01g**. This circuit is a 25:1 step-up circuit; the voltage of the secondary is supposed to be 4 V. Something is wrong. The DMM reading is incorrect. What is the problem? The problem is _____

_____

_____.

Activity 17.2: Transformer Buck/Boost Circuits

1. Transformers can be used to increase or decrease transformer secondary voltages by using additional transformers connected in **buck/boost** circuits to perform that function. For instance, suppose you have a transformer with a secondary voltage of 105 V and you need 120 V; you can connect a transformer (with the same primary voltage) having a secondary voltage of 15 V in a boost connection to provide the additional voltage. If you have a transformer with a secondary of 230 V and you desire 215 V, you can take the same 15 V transformer and connect it in the buck connection to decrease the secondary voltage to the desired potential of 215 V.

2. The trick to this type of connection is to make use of transformer polarity markings (usually dots or numbers on the transformer) to indicate polarity. The primary terminals are to be connected to the same voltage source (observing proper polarity) as displayed in Figure 17-3.

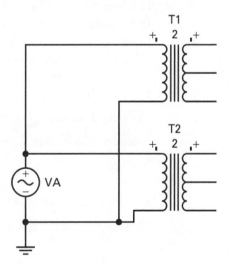

Figure 17-3   Observing Transformer Primary Polarity

3. After the primaries are connected properly, you can proceed with the buck/boost connections of the secondaries (series connections). If the bottom terminal (no dot) of one secondary is connected to the top terminal (dotted) of another secondary, the secondary voltages add (boost).

4. Open circuit file **17-02a**. In this circuit, the primary voltage is 120 V. The secondary voltage should be 240 V, but transformer T_1 is only providing 215 V. A boost transformer (T_2) that provides an additional 25 V is needed. This circuit is connected for that operation to take place. Activate the circuit and use the DMM to measure the secondary voltage of T_1 (TPA to ground) and T_2 (TPA to TPB). What is the total output voltage of the transformer circuit (TPB to ground)? Measured T_1 Secondary Voltage = _____. Measured T_2 Secondary Voltage = _____. Measured Output Voltage = _____.

5. For the buck operation, the secondaries have to be connected with the bottom terminals (no dots) tied together or the top terminals (dots) connected together.

6. Open circuit file **17-02b**. This is the same circuit as the previous circuit (17-02a) but with the secondaries connected in the buck fashion. Activate the circuit and use the DMM to measure the secondary of T_1 (TPA to ground) and T_2 (TPA to TPB). What is the total output of this transformer circuit to (TPB ground)? Measured T_1 Secondary Voltage = _____. Mea- sured T_2 secondary Voltage = _____. Measured Output Voltage = _____.

7. Open circuit file **17-02c**. This circuit is using a typical industrial dual-voltage transformer from the **Electromechanical** menu. Notice that it has only one primary and two secondaries. In this case the primary is 120V. The secondaries are designed to produce one 120-V circuit with increased amperage capabilities (connect X_1 to X_3 and X_2 to X_4), two independent 120-V circuits, or one 240-V circuit. Activate this circuit and measure the secondaries and the output voltage. Measured Secondary 1 (X_1-X_2) Voltage = _____. Measured Secondary 2 (X_3-X_4) Voltage = _____. Measured Output (X_1-X_4) Voltage = _____.

8. Use both secondaries to provide one 120-V output. Measure the output. Measured Output Voltage = _____.

● *Troubleshooting Problems:*

9. Open circuit file **17-02d**. Activate the circuit and measure the output voltage. Measured Output Voltage = _____.

10. The output voltage of this transformer circuit is supposed to be 30 V with transformer T_2 providing a 6-V boost to transformer T_1 (24 V). Identify the problem and repair the circuit. Reactivate the circuit and measure the output voltage. The problem is_____

_____

and I changed   _____

_____

to correct it. Measured Output Voltage = _____.

18. Passive Filter Circuits

References

Electronics Workbench®, *MultiSIM* Version 7

Electronics Workbench®, *MultiSIM* Version 7 User's Guide

Objectives After completing this chapter the student should be able to:

- Operate and analyze passive filter circuits
- Use the MultiSIM Bode plotter to analyze circuits
- Classify passive filters
- Determine response curves for passive filters
- Determine cutoff frequencies for filter circuits
- Analyze coupling and decoupling circuits
- Troubleshoot transformer circuits

Introduction

All reactive components are sensitive to the frequencies of changing signals from signal and power sources as well as AC signals developed in electronic circuits. Passive filters are circuits that use the "frequency sensitive" characteristics of reactive, yet passive components. Some filters will pass certain frequencies, others will block them, some will only pass a band of frequencies, and others will block that band. There are many uses for filters in electronics circuits and you will find them in use throughout the electronics industry.

Filter circuits are categorized in a general manner by their resonance and non-resonance capabilities. Another way that filters are categorized is by the frequency range of their application. Names of filter circuits often reflect their application such as low-pass, high-pass, bandpass, bandstop (notch), and so on.

A special unit of virtual test instrument found only in MultiSIM is known as the **Bode Plotter** (pronounced bow-dee) and can be used to test virtual filter circuits. The **spectrum analyzer**, in conjunction with an oscilloscope, comes closest to doing the same job as a Bode plotter among "real" test instruments. Historically, there have also been some specialized test instruments with built-in oscilloscopes that were used to do similar tasks as the spectrum analyzer, for example, in television bandpass alignment and in specialized military applications.

Activity 18.1: The MultiSIM Bode Plotter

1. The MultiSIM **Bode plotter** (see Figure 18-1) is used to plot or graph the frequency response of a circuit. This X-Y display presents voltage level (gain or magnitude) of the signal being measured on the Y-axis. The Y-axis display is referenced to the signal frequency plotted on the X-axis. This virtual instrument is typically used to determine frequency response and phase shift in AC circuits.

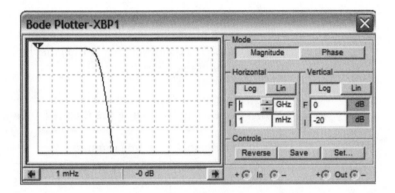

Figure 18-1 The MultiSIM Bode Plotter

2. At the top of the control panel on the top right side of the instrument, the **Magnitude** and **Phase** switches determine the type of measurement to be undertaken. The magnitude setting measures the ratio of magnitudes (voltage gain) between two points (V+ and V−). The phase setting measures the phase shift (in degrees) between two points. Both gain and phase shift are plotted against the signal frequency.

3. On the horizontal axis the display shows the frequency range (in this case, 1 mHz to 1 GHz) being measured. The horizontal axis settings are controlled on the **Horizontal** section of the plotter (Figure 18-2). **I** represents the beginning frequency that will be plotted and **F** represents the ending or final frequency that will be plotted. In the plot for this figure, the sweep will start on the left at 1 mHz and progress to the right side, ending at 1 GHz. These

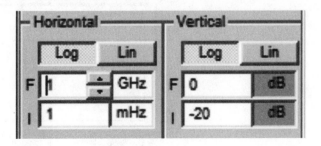

Figure 18-2 Bode Plotter X- and Y-Axis Final and Initial Settings

settings can be changed by clicking on the **GHz** or **mHz**, which brings up a menu that you can change.

4. The units and the scale for the **Vertical Axis** depend on what quantity is being measured and the base being used as shown in Table 18-1. The Y-Axis unit of measurement is the dB (decibel), and in the case of Figure 18-2, displays the voltage output of the circuit from 0 dB to −20 dB with 0 dB being the initial (maximum) output. The output level decreases from this 0 dB maximum to (in this case) −20 dB as frequency changes on the X-Axis. These settings can be changed if you want to look at a specific action of the circuit at a certain area of its response curve. Some circuits will have a rising (right to left) display or a rising and then falling display, depending on the type of circuit being monitored.

When Measuring	Using the Base	Minimum Initial	Maximum Final
Magnitude (gain)	Logarithmic	−200 dB	200 dB
Magnitude (gain)	Linear	0	10e + 09
Phase	Logarithmic	−720°	720°
Phase	Linear	−720°	720°

Table 18-1 Y-Axis Range and Units for the Bode Plotter

5. In a manner similar to the oscilloscope, the Bode plotter has a cursor that can be used to determine voltage and frequency information about specific points on the X-Y display. Figure 18-3 shows the cursor being used to gain specific information; the frequency and voltage information is displayed at the bottom of the plotter screen (in this case, 16.41 Hz and −3.165 dB). The arrows on the left and right sides can be used to move the cursor in intervals. The action is a little rough, but it can help determine information; sometimes

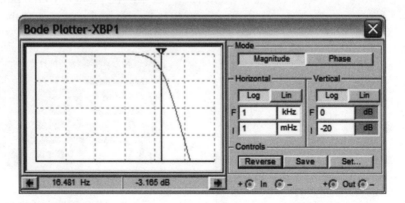

Figure 18-3 Bode Plotter Plot Showing Cursor in Use

you will have to approximate or round off the data (for example, –3.165 dB is about –3 dB).

6. Open circuit file **18-01a** (a low-pass filter). In this circuit, the Bode plotter is set to measure the voltage difference (magnitude) between the signal applied to the circuit and the signal output as seen across the capacitor (TPB). Notice that the Vertical F and I settings are 0 dB and –10 dB, respectively. The display starts at 0 dB at the left of the screen and drops to –10 dB at the right of the screen. Activate the circuit and move the cursor to the right until you get as close to –3 dB as possible (in this case, –3.04 dB). The –3-db point in circuit response is considered to be the **Half-Power** point (or the 0.707 voltage point) as the signal decreases in response to frequency change.

7. The Horizontal Axis **I** (initial) setting for this circuit has been set for 1 Hz and the **F** (final) setting has been set for 200 Hz. The plotter will display circuit action from 1 Hz to 200 Hz. A logarithmic scale is usually used in frequency response analysis on both axes.

8. The frequency response of this circuit at –3 dB is ≈ _____ (the ≈ symbol means approximately in mathematics). Notice that the voltage drop across the capacitor decreases as frequency increases. A signal source (V_A) has to be provided in the circuit for the program to simulate; its voltage level and frequency output has no effect on the Bode plot. See Figure 18-4.

9. Open circuit file **18-01b** (a high-pass filter). In this circuit, an inductor has replaced the capacitor. It is expected that the voltage drop across the inductor will increase as frequency increases, the opposite of the previous circuit.

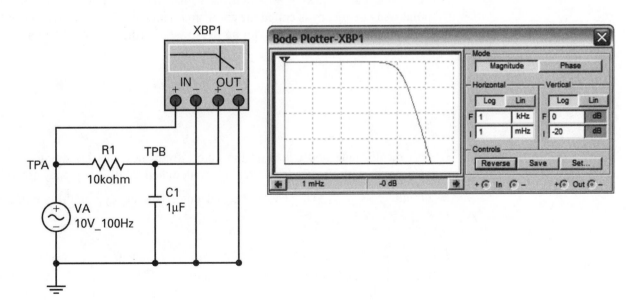

Figure 18-4   A Bode Plot of an RC Circuit

Activate the circuit and observe that the frequency rises from some point out in the frequency spectrum (−100 dB to 0 dB). Move the cursor to the right and determine the frequency at the −3-dB point. The frequency of this circuit at

−3 dB (−2.996 dB) is ≈ _____.

10. Open circuit file **18-01c** (a bandpass filter). In this circuit, the inductor and the capacitor are in parallel with interaction between the reactive components to be expected. Activate the circuit and observe that the frequency rises to some center frequency and falls off past the center frequency point; this response displays the interaction between the active components. Use the cursor to determine the center frequency (maximum voltage level) of this circuit. The center frequency of this circuit at ≈ 0 dB (−.072dB) is

_____. This is also the resonant frequency for this circuit.

Activity 18.2: Low-Pass Filter Circuits

1. A **low-pass filter** circuit passes low frequencies (down to −3 dB) and blocks high frequencies (below −3 dB). The components and their placement in the circuit determine the criterion of what is passed and what is blocked. The −3-dB point on a Bode plot is the half-power point at which higher frequencies are not useable (by definition) as the output of a low-pass filter.

2. Open circuit file **18-02a**. This is a typical low-pass filter circuit similar to the previous circuit (18-01a) of Part 18.1, Step 6. The values of the components have changed and the response will be different. Use the Bode plotter to determine the frequency at which the signal level decreases to the −3 dB point. This frequency is called the cutoff frequency for the filter. The formula to determine cutoff frequency for an RC low-pass filter and an RC high-pass filter is:

$$f_c = \frac{0.159}{RC}$$

3. Use the formula and calculate the cutoff frequency for this circuit. Then measure it with the Bode plotter. The calculated cutoff frequency for this circuit is _____Hz. The measured cutoff frequency for this circuit is _____Hz.

4. To determine phase shift at the cutoff frequency, click on the **Phase** button on the Bode plotter and move the cursor to the cutoff frequency. The amount of phase shift will be displayed to the right of the frequency indication. The phase shift at the cutoff frequency is _____.

5. Open circuit file **18-02b**. This is an RL low-pass filter circuit that acts in a manner similar to the RC low-pass filter. Again, use the Bode plotter to determine the cutoff frequency of this circuit. The formula to determine cutoff frequency for an RL low-pass filter and an RL high-pass filter is:

$$f_c = \frac{0.159R}{L}$$

6. Use the formula and calculate the cutoff frequency for this circuit. Then measure it with the Bode plotter. The calculated cutoff frequency for this circuit is _____Hz. The measured cutoff frequency for this circuit is _____Hz.

7. You can also use the Bode plotter to measure the phase shift of the circuit. What amount of phase shift is present at the cutoff frequency? The amount of phase shift is ≈ _____.

- *Troubleshooting Problem:*

8. Open circuit file **18-02c**. This is a low-pass filter circuit and should have a cutoff frequency of 6.485 kHz, but the output measured by the Bode plotter is about −37 dB. What is wrong with the circuit? Use the DMM to troubleshoot the circuit. The problem with the circuit is that _____

   _____

   _____.

Activity 18.3: High-Pass Filter Circuits

1. A **high-pass filter** circuit blocks lower frequency signals and passes higher frequency signals. As in every filter circuit, the components and their position in the circuit determine the criterion of what is passed and what is blocked. The −3-dB point on a Bode plot is the point at which higher frequencies begin to pass through the filter.

2. Open circuit file **18-03a**. This circuit is an RL high-pass filter that is designed to block low frequencies and pass high frequencies. Calculate the cutoff frequency for this circuit. The calculated cutoff frequency is _____kHz.

3. Activate the circuit and measure the cutoff frequency with the Bode plotter. The measured cutoff frequency is _____kHz.

4. What resistance change could be made to this circuit to lower its cutoff frequency? <u>Increasing/Decreasing</u> (circle one) the resistance value of the resistor would lower cutoff frequency. Change the value of R_1 to the values indicated in Table 18-2, measure the cutoff frequency for each value of R_1, and enter the data in the table.

Change R_1 Value to:	2 kΩ	5 kΩ	7.5 kΩ	10 kΩ	12 kΩ
Measured Cutoff Frequency		11.5 kΩ			

Table 18-2   High-Pass Filter Circuit Cutoff Frequencies

5. Open circuit file **18-03b**. This circuit is an RC high-pass filter. Calculate the cutoff frequency for this circuit. The calculated cutoff frequency is _____Hz.

6. Activate the circuit and measure the cutoff frequency with the Bode plotter. The measured cutoff frequency is _____Hz.

7. What resistance change could be made to this circuit to increase its cutoff frequency? <u>Increasing/Decreasing</u> (circle one) the resistance value of the resistor will increase cutoff frequency. Change the value of R_1 to 1 kΩ and measure the cutoff frequency. The measured cutoff frequency with an R_1 set to 1 kΩ = _____Hz.

● *Troubleshooting Problem:*

8. Open circuit file **18-03c**. This circuit is an RC high-pass filter with a designed cutoff frequency of 265 Hz. There is a problem with the circuit. Activate the circuit and measure the cutoff frequency. Then use the DMM to locate the problem. The measured cutoff frequency is _____kHz. The problem is _____

_____

_____.

Activity 18.4: Band-Pass Filter Circuits

1. A **band-pass filter** circuit blocks low frequency signals up to a certain cutoff point in the frequency spectrum and blocks high frequency signals that are beyond a maximum cutoff frequency. As in every filter circuit, the components and their position in the circuit determines the characteristics of the

circuit. With band-pass filter circuits the concern is with **quality (Q), resonant (center) frequency (f_r), bandwidth**, and the cutoff frequencies (f_c) on either side of the resonance frequency. The sharper the response curve of a resonant circuit, the more selective it is. The less selective a band-pass circuit is, the wider the band of frequencies that is allowed to pass through the circuit, and the wider its bandwidth. **Bandwidth** is the width of the frequency spectrum between the band-pass frequencies at the cutoff points on the frequency plot.

2. The formulas you will need are:

$$Q = \frac{X_L}{R}, \; f_r = \frac{1}{2\pi\sqrt{LC}}, \; XL = 2\pi f_r L, \text{ and } BW = \frac{f_r}{Q}$$

3. Open circuit file **18-04a**. This circuit is a series resonant bandpass filter. At resonance, the inductive and capacitive reactances are equal. Calculate the resonant frequency and bandwidth. Calculated resonant frequency (f_r) is

_____. Calculated bandwidth is _____.

4. Activate the circuit and determine the center frequency and the approximate 70.7% (−3 dB) bandpass points with the Bode plotter. Then calculate bandwidth. Using the Bode plotter, the center frequency is located at

_____kHz. The bandpass is from _____kHz to _____kHz and

bandwidth is _____kHz.

5. Another way to check for bandpass frequencies, bandwidth, and the center frequency is with the function generator and the DMM. Adjust the generator output around the −3-dB points until 7.07 V is read on the DMM on both sides of the response curve. These frequencies are the bandpass points. Watch the voltage reading on the DMM while adjusting the frequency output of the function generator until the center frequency has been ascertained as indicated by a maximum voltage reading. This point might be broad, and it will be necessary to interpolate the center frequency from the extremes. With

the function generator and the DMM, the center frequency is _____kHz.

The bandpass is from _____kHz to _____kHz. The bandwidth is

_____kHz.

6. Open circuit file **18-04b**. This circuit is the same as the one we just worked on except the value of the load resistor R_1 has been changed to 100 Ω. Activate the circuit and notice the changed frequency plot on the Bode plotter. It has a narrow bandwidth compared to the previous circuit. This has occurred because of a change in Q as a result of the change of resistance. Determine the approximate bandpass and bandwidth with the Bode plotter.

The bandpass is from _____kHz to _____kHz. The bandwidth is _____kHz.

7. Open circuit file **18-04c**. This circuit is a parallel resonant bandpass circuit that operates in a similar manner to the series resonant circuit. Activate the circuit and use the Bode plotter to determine the bandpass points, the center frequency, and the bandwidth. Using the Bode plotter, the bandpass points are at _____Hz and _____Hz, the bandwidth is _____Hz, and the center frequency is _____Hz.

● *Troubleshooting Problem:*

8. Open circuit file **18-04d**. This circuit is a series resonant bandpass circuit that is not operating properly. The center frequency is about 1 kHz. At that frequency, the output is unable to reach 0 dB; it is about −6.5 dB. Use the DMM to determine the problem. The problem is _____.

Activity 18.5: Band-Reject, Bandstop, and Notch Filter Circuits

1. Bandstop or band-reject filters (notch filters) prevent certain frequencies from passing through a circuit. Essentially, the methods are the same as for the bandpass filters, but with opposite effects. The center frequency of the notch occurs when the response curve is at its most negative. The band-reject points are the −3-dB points that you determined with bandpass filters. Figure 18-5 displays a typical notch filter output with the Bode plotter cursor at the center frequency of the notch.

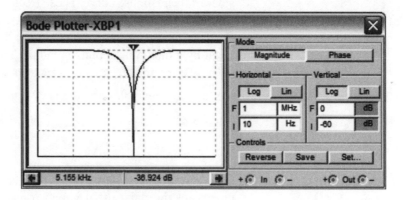

Figure 18-5   Notch Filter Bode Plot

2. Open circuit file **18-05a**. This is a series notch filter. Activate the circuit and measure the band-reject points, the band-reject width, and the center

frequency of the notch with the Bode plotter. Using the Bode plotter, the band-reject points are at _____Hz and _____Hz, the band-reject width is _____Hz, and the center frequency of the notch is _____Hz.

3. Open circuit file **18-05b**. This is a parallel notch filter. Activate the circuit and measure the band-reject points, the band-reject width, and the center frequency of the notch with the Bode plotter. Using the Bode plotter, the band-reject points are at _____Hz and _____Hz, the band-reject width is _____Hz, and the center frequency of the notch is _____Hz.

- ***Troubleshooting Problem:***

4. Open circuit file **18-05c**. This is the same circuit that was worked on in Step 3. There is a problem with this circuit; the notch is poorly defined. Use the DMM and check the components. The problem is _____.